Nuclear Magnetic Resonance

Nuclear Magnetic Resonance

SECOND EDITION

P. J. Hore

OXFORD
UNIVERSITY PRESS

OXFORD

UNIVERSITY PRESS

Great Clarendon Street, Oxford, OX2 6DP,
United Kingdom

Oxford University Press is a department of the University of Oxford.
It furthers the University's objective of excellence in research, scholarship,
and education by publishing worldwide. Oxford is a registered trade mark of
Oxford University Press in the UK and in certain other countries

Published in the United States of America by Oxford University Press
198 Madison Avenue, New York, NY 10016, United States of America

British Library Cataloguing in Publication Data
Data available

Library of Congress Control Number: 2015931304

ISBN 978-0-19-870341-9

Printed in Great Britain by Ashford Colour Ltd, Gosport, Hampshire

Links to third party websites are provided by Oxford in good faith and
for information only. Oxford disclaims any responsibility for the materials
contained in any third party website referenced in this work.

The manufacturer's authorised representative in the EU for product safety is
Oxford University Press España S.A. of El Parque Empresarial San Fernando de
Henares, Avenida de Castilla, 2 – 28830 Madrid (www.oup.es/en or product.
safety@oup.com). OUP España S.A. also acts as importer into
Spain of products made by the manufacturer.

Preface to the 1st edition

Nuclear magnetic resonance is an enormously powerful and versatile technique for investigating the structure and dynamics of molecules. These days it is difficult to find a Chemistry laboratory without an NMR spectrometer or an undergraduate Chemistry course without a set of NMR lectures. This book offers a clear, concise introduction to the physical principles of NMR of liquids, and the interactions that determine the appearance of NMR spectra. The six chapters describe and explain how nuclear spins interact with a magnetic field (the chemical shift) and with each other (spin–spin coupling); how NMR spectra are affected by chemical equilibria (exchange) and molecular motion (relaxation); and conclude with an outline of the workings of a few one- and two-dimensional NMR experiments. I have made every effort to keep things simple: only essential mathematics and theory are included. The emphasis throughout is on understanding the foundations of the technique and how it may be used to study problems of chemical interest.

The shape and content of this short book owe much to those who taught me magnetic resonance—Keith McLauchlan, Rob Kaptein, and Ray Freeman—especially to Keith and Ray whose Oxford undergraduate lecture courses I inherited ten years ago, and on which the book is based. I am indebted and profoundly grateful to all three. My thanks also go to Jonathan Jones, who carefully and perceptively read the whole thing and made penetrating comments on almost every page; to Paul Hodgkinson, Mark Wormald, and Pete Biggs who cheerfully untangled the computer glitches I encountered (and sometimes generated) while drawing the figures; and to Craig Morton and Mark Wormald for generously making available their beautiful spectra.

Oxford, October 1994   P. J. H.

Preface to the 2nd edition

Although the practice of NMR spectroscopy has changed hugely over the last 20 years, the physical principles of liquid-state NMR, with which this little book is concerned, remain essentially the same. The origins of chemical shifts, spin–spin couplings, chemical exchange, and spin relaxation, and their effects on the appearance of NMR spectra, were all already pretty well understood by 1995, at least at the level of most undergraduate Chemistry courses. As a consequence, the ground covered by this second edition does not differ greatly from the first. The most significant additions, aimed at making the coverage of experimental techniques a little more contemporary, are sections on INEPT, HSQC, and three-dimensional NMR.

I have corrected some of the errors, inaccuracies, and oversimplifications in the first edition without, I hope, introducing too many new ones. If you spot anything wrong, please let me know. In retrospect, some parts of the first edition seemed less than satisfactory. In particular, I have reorganized the sections on neighbouring group anisotropy, strong coupling, equivalent spins, dipolar coupling, the vector model, and two-dimensional NMR. The signs of Larmor frequencies and the rotations produced by radiofrequency pulses are now aligned with modern usage. I have also made countless minor changes, intended to make the text easier to follow, and some cosmetic improvements to the figures.

Ten exercises have been added at the end of each chapter. Some are very straightforward, others probably a bit more challenging. I hope they will help you verify and deepen your understanding of the principles of NMR. Answers can be found at the back. Worked solutions are available online at www.oxfordtextbooks. co.uk/orc/hore_nmr2e/. Each chapter also has a short summary section at the end.

My other Oxford Chemistry Primer, *NMR: the Toolkit*, written with Jonathan Jones and Steve Wimperis in 2000, is also going into a second edition. It can be regarded as a continuation of the present volume for those who want to know more about how NMR pulse sequences actually work and to appreciate the underlying quantum mechanics. Readers who are frustrated by the lack of detail and rigour surrounding quantum matters here may find *The Toolkit* more to their taste.

I am grateful to Rod Wasylishen for taking the time to point out some deficiencies in the first edition. Many thanks to Geoffrey Bodenhausen and Gareth Morris who kindly provided detailed comments on the first edition and wise suggestions for improvements. I would also like to thank Susannah Worster for help with the glossary, Jakub Sowa for help with ChemDraw, Christina Redfield for the 950 MHz spectra of lysozyme, and Pete Biggs for resurrecting electronic versions of the figures I drew 20 years ago using a now defunct graphics package.

Oxford, October 2014   P. J. H.

Contents

Introduction

1.1 Introduction

Molecules are inconveniently small and difficult to observe directly. To learn about their structures, motions, reactions, and interactions one needs microscopic spies able to relay information on their molecular hosts without significantly perturbing them. The spies that form the subject of this book are atomic nuclei, and the attribute that makes them successful at espionage is their magnetism.

Nuclear magnetic moments are exquisitely sensitive to their surroundings and yet interact very weakly with them. Most elements have at least one naturally occurring magnetic isotope, so essentially every molecule you can think of has one or more spies already in place, exerting a negligible influence on the molecular properties they probe.

A magnetic nucleus placed in a magnetic field has a small number of quantized energy levels. In the case of the nucleus of a hydrogen atom there are just two. Roughly speaking, they differ in the orientation of the nuclear magnetic moment which can point in the *same* direction as the magnetic field or in the *opposite* direction. The spacing of the two energy levels, ΔE, depends, not surprisingly, on the size of the nuclear magnetic moment and the strength of the magnetic field. ΔE may be measured by using electromagnetic radiation to induce transitions between the two energy levels. This can occur when the frequency, ν_{NMR}, of the applied radiation satisfies the *resonance condition* $\Delta E = h\nu_{NMR}$, where h is Planck's constant. This, essentially, is *nuclear magnetic resonance* (NMR) spectroscopy.

For a given magnetic field strength, the energy gap ΔE and therefore the resonance frequency are determined principally by which nuclide is observed, because every nuclide (^{1}H, ^{2}H, ^{13}C, ^{14}N, ^{15}N, etc.) has a characteristic magnetic moment. But there is far more to NMR than being able to distinguish hydrogen from deuterium, or ^{13}C from ^{14}N. Fortunately for chemists, the resonance frequency also depends, slightly, on the chemical environment of the nucleus in a molecule, an effect known as the *chemical shift*. For example, Fig. 1.1 shows an NMR spectrum of the hydrogen nuclei in liquid ethanol, CH_3CH_2OH. The three different kinds of H atom have different resonance frequencies and so give rise to three separate signals in the spectrum, identifiable by their integrated areas which are in the ratio 3 : 2 : 1, reflecting the number of hydrogens of each type: CH_3, CH_2, OH. This rather low quality spectrum, recorded in the early days of NMR, hints at the way in which NMR can be used to probe molecules at the atomic level.

I am grateful to Ray Freeman for the spy simile.

The elementary introduction to NMR in this chapter is slightly oversimplified in places. This is intentional and is done to avoid a discussion of the quantum mechanics of spins in magnetic fields. A more accurate account of some aspects of spin dynamics can be found in later chapters. For more detailed treatments see, for example, Levitt (2008), Keeler (2010), and Hore, Jones, and Wimperis (2015).

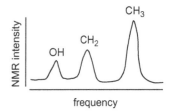

Fig. 1.1 ^{1}H NMR spectrum of liquid ethanol, CH_3CH_2OH, showing the distinct resonance frequencies of the three types of H atom in the molecule. (Adapted from J. T. Arnold, S. S. Dharmatti, and M. E. Packard, *J. Chem. Phys.*, **19** (1951) 507.) A higher-resolution spectrum of ethanol is shown in Fig. 3.1.

1.2 Angular momentum and nuclear magnetism

Having seen very briefly what NMR is, let us go back to the beginning and look at the origins and properties of nuclear magnetism.

Spin angular momentum

Magnetic nuclei possess an intrinsic angular momentum known as *spin*, whose magnitude is quantized in units of $\hbar(=h/2\pi)$:

$$\text{magnitude of spin angular momentum} = \sqrt{I(I+1)}\,\hbar. \tag{1.1}$$

The *spin quantum number, I,* of a nucleus may have one of the following values:

$$I = 0, \tfrac{1}{2}, 1, \tfrac{3}{2}, 2, ..., \tag{1.2}$$

with $I > \tfrac{7}{2}$ being rather rare. Spin quantum numbers for a selection of nuclei are given in Table 1.1. Notice that isotopes of the same element may have different quantum numbers and that some common nuclei, notably ^{12}C and ^{16}O, have $I = 0$, i.e. no angular momentum, no magnetic moment, and consequently no NMR spectra. A nucleus that has spin quantum number I is said to be or have 'spin-I'. Electrons, neutrons, and protons (^1H nuclei) have spin-$\tfrac{1}{2}$. Following common practice in NMR, we use 'spins' as a synonym for magnetic nuclei, and 'protons' as a synonym for ^1H nuclei.

The spin quantum number of a nucleus is determined largely by the number of unpaired nucleons. For example, an isotope such as ^{12}C has even numbers of protons and neutrons: all the protons pair up with antiparallel spins, as do all the neutrons, giving a net spin angular momentum of zero ($I = 0$). A nucleus with odd numbers of protons and neutrons (e.g. ^{14}N, seven of each) generally has an integral, non-zero quantum number because the total number of unpaired nucleons is even, and each of them contributes $\tfrac{1}{2}$ to I. However, it is difficult to predict exactly how many protons and neutrons will be unpaired except in simple cases such as ^2H. These ideas can be extended to nuclei with even numbers of protons and odd numbers of neutrons, or vice versa, which usually have a half-integral quantum number due to an odd number of unpaired nucleons. These rules of thumb, which are not infallible, are collected in Table 1.2.

Spin angular momentum can be thought of as a vector, $\boldsymbol{I}$, whose *direction* and *magnitude* are quantized. In this vector representation, $\boldsymbol{I}$ has length $\sqrt{I(I+1)}\,\hbar$ (eqn 1.1) and has $2I + 1$ allowed projections onto an arbitrarily chosen axis, labelled z. That is, the projection of $\boldsymbol{I}$ onto the z-axis is quantized:

$$I_z = m\hbar, \tag{1.3}$$

where m, the magnetic quantum number, has $2I + 1$ values between $+I$ and $-I$:

$$m = I, I-1, I-2, ... -I. \tag{1.4}$$

Table 1.1 Nuclear spin quantum numbers (I) of some commonly occurring nuclides

I	Nuclide
0	$^{12}C, ^{16}O$
$\tfrac{1}{2}$	$^1H, ^{13}C, ^{15}N, ^{19}F, ^{29}Si, ^{31}P$
1	$^2H, ^{14}N$
$\tfrac{3}{2}$	$^{11}B, ^{23}Na, ^{35}Cl, ^{37}Cl$
$\tfrac{5}{2}$	$^{17}O, ^{27}Al$
3	^{10}B

Vectors (quantities that have direction as well as magnitude) are printed in bold, italic type. The spin angular momentum vector, $\boldsymbol{I}$, should not be confused with the spin quantum number, I.

Table 1.2 Rules for predicting nuclear spin quantum numbers (I) from the numbers of protons and neutrons in a nucleus

Number of protons	Number of neutrons	I
even	even	0
odd	odd	1 or 2 or 3 or . . .
even	odd	$\frac{1}{2}$ or $\frac{3}{2}$ or $\frac{5}{2}$ or . . .
odd	even	$\frac{1}{2}$ or $\frac{3}{2}$ or $\frac{5}{2}$ or . . .

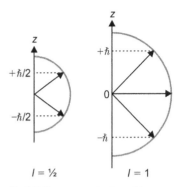

Fig. 1.2 Space quantization of the angular momentum of spin-$\frac{1}{2}$ and spin-1 nuclei. The lengths of the vectors (eqn 1.1) are $\frac{\sqrt{3}}{2}\hbar$ and $\sqrt{2}\hbar$, respectively. Their z-components are given by eqns 1.3 and 1.4.

For example, the angular momentum of a spin-$\frac{1}{2}$ nucleus (e.g. ^{1}H, ^{13}C) has two permitted z-components, $I_z = \pm\frac{1}{2}\hbar$, while a nucleus with $I = 1$ has three, $I_z = 0, \pm\hbar$. This *space quantization* is illustrated in Fig. 1.2. In the absence of a magnetic field, the axis of quantization is arbitrary and all $2I + 1$ directions are equally likely and have the same energy.

Nuclear magnetic moments

The magnetic moment of a nucleus is intimately connected with its spin. To be more precise, the magnetic moment $\boldsymbol{\mu}$ (also a vector quantity) is related to $\boldsymbol{I}$ by:

$$\boldsymbol{\mu} = \gamma \boldsymbol{I}, \tag{1.5}$$

The magnitude of the magnetic moment, i.e. the length of the vector $\boldsymbol{\mu}$, is $\gamma\sqrt{I(I+1)}\,\hbar$.

where γ is the magnetogyric ratio (or gyromagnetic ratio) of the nucleus.

The magnetogyric ratios of some common NMR nuclei are given in Table 1.3. Note that the magnetic moment of a nucleus is *not* simply the sum of the magnetic moments of the unpaired protons and neutrons.

To summarize, the magnetic moment of a nucleus is parallel to the spin angular momentum or, sometimes, antiparallel for nuclei that have negative γ. The magnitude and orientation of both are quantized.

Table 1.3 Magnetogyric ratios, NMR frequencies (in a 9.4 T field), and natural abundances of selected nuclides

	$\gamma/10^7\ \text{T}^{-1}\ \text{s}^{-1}$	ν_{NMR}/MHz	Natural abundance/%
^{1}H	26.752	400.0	99.985
^{2}H	4.107	61.4	0.015
^{13}C	6.728	100.6	1.108
^{14}N	1.934	28.9	99.63
^{15}N	−2.713	40.5	0.37
^{17}O	−3.628	54.3	0.037
^{19}F	25.181	376.5	100.0
^{29}Si	−5.319	79.6	4.70
^{31}P	10.839	162.1	100.0

The scalar product of two vectors a and b is $a \cdot b = ab\cos\theta$, where θ is the angle between them and a and b are their magnitudes. The projection of b onto a has magnitude $b\cos\theta$.

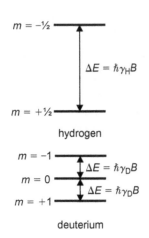

hydrogen

deuterium

Fig. 1.3 Energy levels for hydrogen ($I = \frac{1}{2}$) and deuterium ($I = 1$) nuclei in a magnetic field B. Note that $\gamma_H > \gamma_D > 0$. The energy-level splittings are not drawn to scale. For nuclei with $\gamma < 0$ (e.g. ^{15}N), the level with the most negative m has the lowest energy.

γ appears in eqns 1.9 and 1.10 as $|\gamma|$ to ensure that ΔE and v_{NMR} are both positive for nuclei with negative magnetogyric ratios, e.g. ^{15}N and ^{29}Si.

Energy levels

In the absence of a magnetic field, the $2I + 1$ energy levels of a spin-I nucleus have the same energy. This degeneracy is removed when a magnetic field is applied: classically, the energy of a magnetic moment μ in a magnetic field B (yet another vector) can be written in terms of the scalar product of the two vectors:

$$E = -\mu \cdot B. \tag{1.6}$$

In the presence of a strong field the spin quantization axis (z) is no longer arbitrary but coincides with the field direction. Therefore:

$$E = -\mu \cdot B = -\mu_z B, \tag{1.7}$$

where μ_z is the z-component of μ (the projection of μ onto B) and B is the strength of the field (the magnitude of the vector B). From eqns 1.3 and 1.5, $\mu_z = \gamma I_z = m\hbar\gamma$ so that

$$E = -m\hbar\gamma B. \tag{1.8}$$

That is, the energy of the nucleus is shifted by an amount proportional to the magnetic field strength, the magnetogyric ratio, and the z-component of the angular momentum. The $2I + 1$ energy levels of a spin-I nucleus are thus equally spaced, with an energy gap $\hbar\gamma B$ (Fig. 1.3).

The selection rule for NMR spectroscopy is $\Delta m = \pm 1$ so that the allowed transitions occur between adjacent energy levels. Therefore the resonance condition is

$$\Delta E = \hbar|\gamma|B \tag{1.9}$$

and the resonance frequency ($= \Delta E/h$) is

$$v_{NMR} = \frac{|\gamma|B}{2\pi}. \tag{1.10}$$

All $2I$ allowed transitions of a spin-I nucleus have the same energy. More on quantization of angular momentum may be found in Atkins and Friedman (2011).

The magnetic field experienced by a nucleus *in an atom or a molecule* differs slightly from the external field such that the exact resonance frequency is characteristic of the chemical environment of the nucleus. This is the origin of the chemical shift (Fig. 1.1, Section 1.3, and Chapter 2).

1.3 NMR spectroscopy

Resonance frequencies and chemical shifts

A typical magnetic field strength used for NMR is 9.4 T (T = tesla), roughly 10^5 times stronger than the Earth's magnetic field. For hydrogen nuclei, eqn 1.10 predicts a resonance frequency $v_{NMR} = 4 \times 10^8$ Hz = 400 MHz (see Table 1.3 for the value of γ). This falls in the radiofrequency region of the electromagnetic spectrum and corresponds to a wavelength of 75 cm. The radiation required to

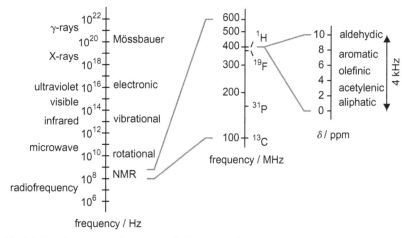

Fig. 1.4 The electromagnetic spectrum (left). Expanded regions show NMR frequencies of nuclei in a 9.4 T field (middle), and typical ^{1}H chemical shifts in parts per million, ppm (right).

induce NMR transitions is consequently referred to as the *radiofrequency field*. Magnetic fields in the range 2.35–23.5 T are commonly used, giving ^{1}H resonance frequencies of 100–1000 MHz. Since ^{1}H is by far the most popular NMR nucleus (for reasons that will soon emerge), NMR spectrometers are usually classified by their ^{1}H frequencies, rather than the strengths of their magnetic fields. Table 1.3 summarizes the NMR properties of several magnetic nuclei: magnetogyric ratio, resonance frequency in a 400 MHz spectrometer, and natural isotopic abundance. Of these nuclei, ^{1}H has the largest magnetogyric ratio (and the largest magnetic moment); in fact the γ of ^{1}H is exceeded only by that of the radioactive isotope tritium, ^{3}H.

To put NMR into context, Fig. 1.4 shows the electromagnetic spectrum, on a logarithmic scale, from the radiofrequency region ($\sim 10^6$ Hz) through microwaves, infrared, visible, and ultraviolet to X-rays and γ-rays ($\sim 10^{22}$ Hz). NMR lies at the low frequency end of the spectrum; most other spectroscopies—rotational, vibrational, electronic, Mössbauer—are concerned with larger energy-level spacings and hence higher frequencies. Also shown, on an expanded logarithmic scale, is the spectral region between 100 and 600 MHz, with the resonance frequencies of a few nuclides in a 9.4 T field. In addition, as a foretaste of the next chapter, a narrow slice of frequencies close to 400 MHz is expanded to show some typical ^{1}H chemical shifts for a few organic functional groups. Notice that the total width of this part of the figure is only 4 kHz. Although other nuclei have somewhat larger chemical shifts, the differences in resonance frequencies between nuclides almost always exceeds their chemical shift ranges, so that it is rare to find overlapping spectra from different nuclides.

Populations and polarizations

At thermal equilibrium, the populations of the energy levels of identical, non-interacting spins in a magnetic field are given by the Boltzmann distribution.

Consider ^{1}H nuclei in a 9.4 T field at a temperature T. Using α and β to label the lower ($m=+\frac{1}{2}$) and upper ($m=-\frac{1}{2}$) energy levels, respectively, the ratio of the populations is

$$\frac{n_\beta}{n_\alpha} = \exp(-\Delta E/k_B T), \tag{1.11}$$

where $\Delta E = \hbar \gamma_H B$ (eqn 1.9) and k_B is Boltzmann's constant. Evaluating this expression at 300 K, one finds $\Delta E = 2.65 \times 10^{-25}$ J, $k_B T = 4.14 \times 10^{-21}$ J, and $\Delta E/k_B T = 6.4 \times 10^{-5}$. Thus, the energy required to reorient the spins is dwarfed by the thermal energy $k_B T$, so that there will be only a small excess of spins in the lower energy level. For such small values of $\Delta E/k_B T$, eqn 1.11 may be simplified using $\exp(-x) \approx 1 - x$, to obtain the relative population difference, or polarization, p

$$p = \frac{n_\alpha - n_\beta}{n_\alpha + n_\beta} \approx \frac{\Delta E}{2k_B T}. \tag{1.12}$$

With the above numbers, eqn 1.12 gives $p \approx 3.200 \times 10^{-5}$. That is, for every 15,624 spins in the upper level there are 15,625 in the lower level. The polarization will be even smaller for ^{1}H nuclei in a weaker field, or for spins with smaller $|\gamma|$ (i.e. almost all other nuclei). This situation is in stark contrast to electronic spectroscopy at a frequency of, say, 10^{15} Hz. Here, ΔE is much larger than $k_B T$ at room temperature and almost all of the molecules are in the ground state, leaving the excited state virtually empty.

In any form of spectroscopy, an electromagnetic field excites molecules or atoms or electrons or nuclei as the case may be, from the lower energy level to the upper one with the *same* probability as it induces the reverse transition, from excited state to ground state. The net absorption of energy, and hence the intensity of the spectroscopic transition, therefore depends on the *difference* in populations of the two levels. In NMR spectroscopy, where the upward transitions outnumber the downward transitions by only one part in 10^4–10^6, it is as if one *detects* only one nucleus in every 10^4–10^6. Add to this the fact that spectroscopy at higher frequencies is much more sensitive as a rule, because higher energy photons are easier to detect, and it becomes clear that NMR signals must be rather weak. It is therefore of crucial importance to optimize signal strengths, for example by using strong magnetic fields to maximize ΔE and therefore the polarization. Similarly, nuclei with large magnetogyric ratio and high natural abundance are favoured (Table 1.3), hence the popularity of ^{1}H as an NMR nucleus.

Spin–spin coupling

The chemical shift (Chapter 2) is not the only source of information encoded in liquid-state NMR spectra. Magnetic interactions between nuclei lead to extra NMR lines which give valuable clues to the arrangement of atoms in molecules.

For example, the spectrum of ethanol in Fig. 1.1 in fact comprises eight distinct peaks (Chapter 3) and sometimes more (Chapter 4) when recorded on a modern spectrometer with less severe instrumental line broadening. For NMR of molecules in the solid state, additional spectral features come from *dipolar* spin–spin interactions (Chapter 3 and Appendix A).

Linewidths

The NMR lines of many nuclei are exceedingly narrow: it is relatively straightforward on a modern spectrometer to resolve two resonances that differ in frequency by only one part in 10^9 (linewidths in ^{1}H spectra of small molecules in solution can be as small as ~0.1 Hz). This quite staggering resolution comes about because nuclear magnetic moments are feeble, and interact very weakly with their surroundings (Chapter 5).

But not all nuclei give sharp lines. For reasons discussed later (Chapter 5), the NMR lines of nuclei with spin quantum numbers greater than $\frac{1}{2}$ are often broad—100 Hz or more. The majority of NMR experiments are therefore performed on spin-$\frac{1}{2}$ nuclei, the most popular being ^{1}H, ^{13}C, ^{19}F, and ^{31}P.

NMR lines can also be broadened by chemical and conformational equilibria (Chapter 4).

Experimental methods

There are three essential requirements for an NMR experiment: a strong magnetic field to polarize the spins, radiofrequency radiation to excite them, and equipment to detect the resulting NMR signal. Throughout this book we concentrate on NMR of molecules in the *liquid* state. See Duer (2004) and Apperley, Harris and Hodgkinson (2012) for solid-state NMR.

The magnetic field is most commonly provided by a superconducting solenoid—a coil of resistance-free alloy supporting a persistent current. At the time of writing, NMR magnets up to 23.5 T (1000 MHz ^{1}H frequency) are commercially available. At the centre of the magnet sits the NMR 'sample', typically ~1 ml of a solution of the molecule of interest contained in a ~5 mm diameter, cylindrical glass tube.

To observe magnetic resonance, one needs coherent, monochromatic electromagnetic radiation at the appropriate radiofrequency. This is generated by passing a sinusoidally oscillating current through a coil of wire or foil arranged around the sample. The electrons circulating in the coil generate an oscillating electromagnetic field whose magnetic component excites the spins. As a consequence, the net magnetization of the sample rotates around the static field and induces an alternating current in the coil which ultimately leads to the NMR signal.

In the early days of NMR (1945–1970), experiments were carried out almost exclusively using *continuous wave* methods, with the radiofrequency field present throughout. Either one kept the radiofrequency fixed, while varying the magnetic field strength, or vice versa, so as to bring nuclei with different chemical shifts sequentially into resonance. Since 1970, *pulse* methods have almost completely taken over. A short, intense burst of radiofrequency radiation (a 'pulse') is applied to the sample to produce an oscillating magnetization that is detected and processed in a computer to give the spectrum. As outlined in Chapter 6, pulse methods have overwhelming advantages over the continuous wave approach.

1.4 Summary

- Every nuclide has a characteristic spin quantum number, I.
- Every nuclide with $I \neq 0$ has a characteristic magnetic moment.
- Nuclear spins have $2I + 1$ energy levels.
- In a magnetic field, the energy levels are split apart by an amount proportional to the size of the nuclear magnetic moment and the strength of the field.
- NMR spectroscopy uses electromagnetic radiation to cause transitions between the energy levels.
- The frequencies of NMR transitions are typically in the range 10–1000 MHz.
- The intensities of NMR signals depend on the differences in the populations of the energy levels.

1.5 Exercises

Answers to the exercises are provided at the back of the book. Full worked solutions are available on the Online Resource Centre at www.oxfordtextbooks.co.uk/orc/hore_nmr2e/

Nuclear spin quantum numbers and magnetogyric ratios can be found in Tables 1.1 and 1.3.

1. Two of the following nuclides have spin quantum number $I = 0$, two have $I = \frac{1}{2}$, and two have $I = 1$: ^6_3Li, $^{14}_7N$, $^{32}_{16}S$, $^{40}_{20}Ca$, $^{57}_{26}Fe$, $^{119}_{50}Sn$. Which is which?

2. The spin angular momentum of ^{23}Na is 2.042×10^{-34} J s. What is its spin quantum number?

3. Calculate the magnetic moment of a ^{14}N nucleus.

4. What are the NMR frequencies (ν_{NMR}) of (a) ^1H, (b) ^{13}C, and (c) ^{15}N in a 23.488 T magnetic field?

5. What is the NMR frequency of ^1H in the Earth's magnetic field (50 μT)? Why is a 50 μT magnetic field rarely used for NMR experiments?

6. The NMR frequency of a nuclide in a 17.6 T magnetic field is 76 MHz. Identify the nuclide.

7. Confirm that eqn 1.11 leads to eqn 1.12 in the limit $\Delta E \ll k_B T$.

8. What is the polarization, p, of ^{1}H nuclei in a 17.6 T magnetic field at 300 K?

9. (a) The polarization, p, of ^{1}H nuclei in a magnetic field at 300 K is 0.01. What is the ^{1}H resonance frequency in this (unrealistically strong) field? (b) What temperature would be required to get the same ^{1}H polarization ($p = 0.01$) using a 400 MHz spectrometer?

10. Suppose it were possible to transfer all of the polarization of electron spins in a 4.0 T magnetic field at 4 K to protons in a 600 MHz spectrometer at 300 K. By what factor would the ^{1}H NMR signal be enhanced? Electrons have spin-$\frac{1}{2}$ and magnetogyric ratio $\gamma_e = -1.761 \times 10^{11}$ T^{-1} s^{-1}. Note that eqn 1.12 is not valid for electron spins under the above conditions.

2 Chemical shifts

2.1 Introduction

The NMR frequency of a nucleus in a molecule is determined principally by its magnetogyric ratio, γ, and the strength, B, of the magnetic field it experiences (eqn 1.10):

$$\nu_{NMR} = \frac{|\gamma|B}{2\pi}.$$

(2.1)

Thus, ^{1}H and ^{13}C nuclei resonate, respectively, at 400 and 100.6 MHz in a 9.4 T field. But not all protons, nor all carbons, have identical resonance frequencies: ν_{NMR} depends (slightly) on the position of the nucleus in the molecule, or to be more precise, on the local electron distribution. This effect, the *chemical shift*, is one of the things that make NMR so attractive to chemists. It makes it possible to distinguish, for example, the three kinds of hydrogen atoms in ethanol (Fig. 1.1) and gives separately detectable signals for the hundreds of hydrogen atoms in a protein (Fig. 2.1).

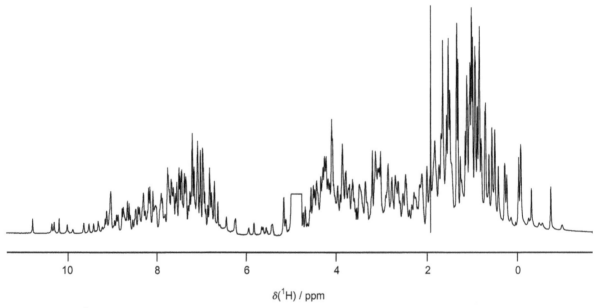

Fig. 2.1 950 MHz ^{1}H spectrum of hen egg-white lysozyme in H_2O. This protein has 129 amino acid residues and a relative molecular mass of ~14,500. The resonance of the solvent protons at ~4.7 ppm has been truncated. This spectrum was kindly provided by C. Redfield.

2.2 Nuclear shielding

Chemical shifts arise because the field, B, actually experienced by a nucleus in an atom or molecule differs slightly from the external field, B_0, produced by the magnet. B_0 is the field that would be felt by a bare nucleus, stripped of its electrons. In an atom, B is slightly smaller than B_0 because the external field causes the electrons to circulate within their atomic orbitals; this induced motion, much like an electric current passing through a coil of wire, generates a small magnetic field B' in the *opposite* direction to B_0 (Fig. 2.2). The nucleus is thus said to be *shielded* from the external field by its surrounding electrons ($B = B_0 - B'$).

B' is proportional to B_0 (the stronger the external field, the more it 'stirs up' the electrons) and typically $10^4 - 10^5$ times smaller. Thus, the field at the nucleus may be written

$$B = B_0(1-\sigma),\tag{2.2}$$

where σ, the constant of proportionality between B' and B_0, is called the *shielding constant* or *screening constant*. As a result of nuclear shielding, the resonance condition (eqn 2.1) becomes

$$\nu_{\text{NMR}} = \frac{|\gamma| B_0}{2\pi}(1-\sigma),\tag{2.3}$$

i.e. the resonance frequency of a nucleus in an atom is slightly lower than that of a bare nucleus, stripped of all its electrons (Fig. 2.3).

Similar effects occur for nuclei in molecules, except that the motion of the electrons is rather more complicated than in atoms with the result that the induced fields may *augment* or *oppose* the external field. Nevertheless, the effect is still referred to as nuclear shielding. Both the size and sign of the shielding constant in eqn 2.3 are determined by the electronic structure of the molecule in the vicinity of the nucleus. The resonance frequency of a nucleus is therefore characteristic of its environment.

Larmor frequency

As mentioned briefly at the end of Chapter 1, the signal in an NMR experiment arises from the motion of the magnetization of the sample in the strong magnetic field of the spectrometer. Consider a collection of identical, non-interacting nuclear spins experiencing a magnetic field **B**. The *magnetization vector* **M** (the sum of the magnetic moment vectors of the individual spins) moves as shown in Fig. 2.4. It *precesses* around **B**, maintaining a constant angle with respect to **B** and therefore a constant projection onto **B**. This motion, akin to that of the axis of a spinning top or gyroscope, is known as *Larmor precession*. The frequency of this motion, known as the *Larmor frequency*, is given by

$$\nu_0 = -\frac{\gamma B_0}{2\pi}(1-\sigma).\tag{2.4}$$

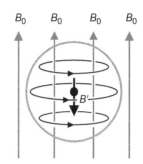

Fig. 2.2 An applied magnetic field B_0 causes the electrons in an atom to circulate within their orbitals. This motion generates an extra field B' which opposes B_0 and results in a net field $B = B_0 - B'$ at the site of the nucleus.

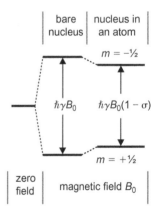

Fig. 2.3 Energy levels of a spin-$\frac{1}{2}$ nucleus with $\gamma > 0$. The energy-level splitting for a nucleus in an atom equals $h\nu_{\text{NMR}}$ where ν_{NMR} is given by eqn 2.3.

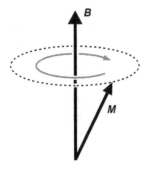

Fig. 2.4 Larmor precession of the magnetization vector M of nuclear spins with a positive magnetogyric ratio ($\gamma > 0$) and hence negative Larmor frequency ($\nu_0 < 0$) in a magnetic field B. When $\gamma < 0$, M precesses in the opposite sense.

ν_0 is negative for nuclei with $\gamma > 0$, e.g. ^{1}H and ^{13}C, and positive for those with $\gamma < 0$, e.g. ^{15}N and ^{29}Si (Table 1.3). The sign of ν_0 determines the sense of the Larmor precession (clockwise or anticlockwise), as indicated in Fig. 2.4. From eqns 2.3 and 2.4, $\nu_0 = -\nu_{NMR}$ for spins with positive γ and $\nu_0 = \nu_{NMR}$ for those with negative γ. Remember that ν_{NMR} was defined (eqn 1.10) as the energy-level spacing ΔE divided by Planck's constant and is always positive. Larmor precession is discussed in more detail in Chapter 6.

Throughout this book, the Greek letter *nu* (e.g. in ν_0 and ν_{NMR}) is used as the symbol for frequencies expressed in hertz (Hz) or equivalently cycles per second or simply s^{-1}. From time to time we will also use the Greek letter *omega* (ω) for *angular* frequencies which have dimensions of radians per second (rad s^{-1}). The two types of frequency differ by a factor of 2π. So, for example, the Larmor angular frequency is given by

$$\omega_0 = 2\pi\nu_0 = -\gamma B_0(1-\sigma). \tag{2.5}$$

Defining chemical shifts

The shielding constant σ is an inconvenient measure of the chemical shift. Since absolute shifts are rarely needed and difficult to determine, it is common practice to define the chemical shift in terms of the *difference* between the Larmor frequency of the nucleus of interest (ν_0) and that of a reference nucleus ($\nu_{0,ref}$) using the dimensionless parameter δ:

$$\delta = 10^6\left(\frac{\nu_0 - \nu_{0,ref}}{\nu_{0,ref}}\right). \tag{2.6}$$

The frequency difference $\nu_0 - \nu_{0,ref}$ is divided by $\nu_{0,ref}$ so that δ is independent of the strength of the magnetic field used to measure it. The factor of 10^6 simply scales the numerical value of δ to a more convenient size: δ values are quoted in *parts per million*, or ppm.

To see how δ is related to the shielding constants, eqns 2.4 and 2.6 can be combined to give

$$\delta = 10^6\left(\frac{\sigma_{ref} - \sigma}{1-\sigma_{ref}}\right) \approx 10^6(\sigma_{ref} - \sigma), \tag{2.7}$$

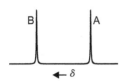

Fig. 2.5 NMR spectra are conventionally plotted with the chemical shift δ increasing from right to left. Spin A is more strongly shielded (larger σ) than spin B and so appears to the right of B in the spectrum. For spins with $\gamma > 0$, A has a less negative, i.e. larger, ν_0 than B.

where $\sigma_{ref} \ll 1$ has been used. The larger the value of σ (greater shielding) the smaller the chemical shift. δ is therefore a *deshielding* parameter.

The reference signal is most conveniently obtained by adding a small amount of a suitable compound to the NMR sample. For ^{1}H and ^{13}C spectra this is usually tetramethylsilane ($CH_3)_4Si$, known as TMS. This molecule is inert, soluble in most organic solvents, and gives a single, strong ^{1}H resonance from its 12 identical protons. Moreover, both the ^{1}H and ^{13}C nuclei in TMS are quite strongly shielded (large shielding constant), with the result that most ^{1}H and ^{13}C chemical shifts are positive numbers ($\delta > 0$).

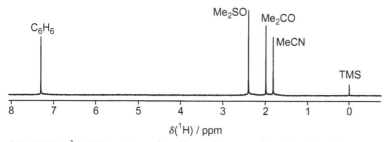

Fig. 2.6 400 MHz ^{1}H NMR spectrum of a mixture of benzene, dimethyl sulphoxide, acetone, acetonitrile, and tetramethylsilane.

Conventionally, NMR spectra are plotted with δ increasing from right to left. Thus, more heavily shielded nuclei (larger σ, smaller δ) appear towards the right-hand side of the spectrum (Fig. 2.5).

As a simple example, Fig. 2.6 shows the 400 MHz ^{1}H spectrum of a mixture of compounds with a small amount of TMS added. Each of the five molecules has a single group of identical protons, and hence a single chemical shift. Notice that the chemical shift scale covers about 10 ppm, a typical range for ^{1}H. Chemical shifts can easily be converted back into frequencies using eqn 2.6. For example, the acetone peak in Fig. 2.6 has $\delta = 2.0$ ppm, so that

$$\left|v_{0,\,acetone} - v_{0,\,TMS}\right| = \left(\delta / 10^6\right)\left|v_{0,\,TMS}\right|$$
$$= \left(2.0\times10^{-6}\right)\times\left(400\,\text{MHz}\right) = 800\,\text{Hz}. \tag{2.8}$$

On a 100 MHz spectrometer, the chemical shift of acetone is still 2.0 ppm, but the Larmor frequency relative to TMS is reduced proportionately to 200 Hz.

Examples

As we shall see later, so many factors play a role in determining the size of chemical shifts that it is often difficult to relate experimental measurements *quantitatively* to molecular structure. However, valuable information can often be extracted from NMR spectra simply by noting the number of resonances and their relative intensities, as the examples in the following paragraphs illustrate.

In CS$_2$ solution, phosphorus pentachloride has a single ^{31}P resonance (Fig. 2.7), as might be expected from the trigonal bipyramidal structure found in the gas phase. Solid phosphorus pentachloride, however, has *two* equally intense ^{31}P peaks, revealing clearly that the change of phase is accompanied by a change in structure (actually the disproportionation reaction $2PCl_5 \rightarrow PCl_6^- + PCl_4^+$).

The ^{17}O NMR spectrum of Co$_4$(CO)$_{12}$ in chloroform at low temperature comprises four equally intense lines (Fig. 2.8), consistent with a bridged structure (C_{3v} symmetry) containing four distinct types of carbonyl. This spectrum clearly rules out a non-bridged Ir$_4$(CO)$_{12}$-type structure (T_d symmetry), in which all 12 carbonyls are in identical environments. It also provides strong evidence against the D_{2d} structure at one time proposed for Co$_4$(CO)$_{12}$ in which there are only *three* distinct carbonyl environments: one bridging and two terminal.

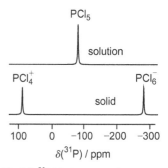

Fig. 2.7 ^{31}P NMR spectra of phosphorus pentachloride in the solid state and in solution in CS$_2$. The former was obtained using a technique known as *magic angle spinning* to remove the large line broadening caused by, amongst other things, the dipolar interactions (Section 3.8 and Appendix A) between nuclei in the solid. The chemical shift reference is an 85% aqueous solution of orthophosphoric acid. (Adapted from E. R. Andrew, *Phil. Trans. R. Soc. A*, **299** (1981) 505.)

Figures 2.7–2.9 and several others in subsequent chapters show NMR spectra that have been drawn on a computer so as to resemble the original experimental spectra (references to which are normally given in the figure captions). Genuine spectra, e.g. Figs 2.6, 2.10, and 2.11, can be recognized as such by the presence of noise.

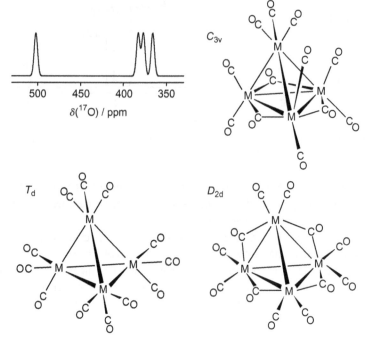

Fig. 2.8 The ^{17}O NMR spectrum of $Co_4(CO)_{12}$ in chloroform at −25 °C is consistent with the C_{3v} structure shown, but not the T_d or D_{2d} forms.

Occasionally, structural information can be deduced solely from the relative intensities of NMR lines. For example, the only possible structures for the compounds C_9H_{12} and $C_{10}H_{14}$ with the ^1H spectra shown in Fig. 2.9 are 1,3,5-trimethylbenzene and 1,2,4,5-tetramethylbenzene, respectively.

A dramatic illustration of the use of chemical shifts is provided by the ^{13}C NMR spectrum of the fullerene C_{60} (Fig. 2.10). The observation of a single NMR line for this remarkable molecule provides direct evidence for its highly symmetrical football-like structure in which all 60 carbon atoms are in identical environments.

Zeolites are aluminosilicates built from corner-sharing SiO_4 and AlO_4 tetrahedra. Each silicate and aluminate group is linked, via oxygen bridges, to four

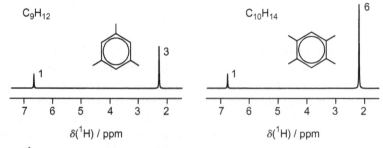

Fig. 2.9 ^1H spectra of compounds with molecular formulae C_9H_{12} and $C_{10}H_{14}$. The relative intensities of the peaks, obtained by integration, are as shown.

other tetrahedra to give framework structures containing cavities and channels, which confer useful catalytic properties. Up to five distinct chemical shifts can be observed in the ^{29}Si NMR spectra of powdered zeolites (Fig. 2.11), corresponding to Si atoms linked to n AlO$_4$ tetrahedra and $(4 - n)$ SiO$_4$ tetrahedra, with $n = 0-4$. Each Al atom shifts the Si resonance by roughly +5 ppm. The relative intensities of the five peaks give the Si/Al ratio, and can be used to test model structures with different Si/Al ordering patterns.

Chemical shifts may be interpreted empirically using data derived from compounds of known structure. For example, Fig. 2.12 shows typical ^{1}H chemical shift ranges for assorted organic functional groups. When combined with empirical rules for predicting substituent effects (see e.g. Friebolin (2011), Günther (2013), and Williams and Fleming (2007)), such tables can be extremely useful in making connections between observed shifts and molecular structures. However, as NMR techniques become ever more sophisticated (Chapter 6), such methods generally become less important. More direct structural information is often provided by spin–spin couplings (Chapter 3) and nuclear Overhauser effects (Chapter 5).

Finally, a somewhat different use of chemical shifts is illustrated by the pH dependence of the ^{1}H spectrum of the amino acid histidine (Fig. 2.13). The resonance frequencies of the H2 and H4 protons in the imidazole group change smoothly with pH between the chemical shifts of the charged form HisH$^+$, stable in acidic solution, and those of the neutral, deprotonated form His, which is present at high pH. At any pH, the observed chemical shift is a weighted average of the two extreme values δ_{HisH^+} and δ_{His}:

$$\delta = \frac{\delta_{HisH^+}[HisH^+] + \delta_{His}[His]}{[HisH^+] + [His]},$$

where [HisH$^+$] and [His] are the concentrations. The details of this averaging process are discussed in Chapter 4. The midpoint of the titration occurs when

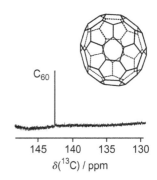

Fig. 2.10 ^{13}C NMR spectrum of C$_{60}$. (Adapted from R. Taylor, J. P. Hare, A. K. Abdul-Sada, and H. W. Kroto, *J. Chem. Soc. Chem. Commun.* (1990) 1423.)

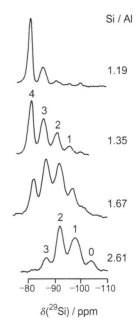

Fig. 2.11 ^{29}Si NMR spectra of synthetic zeolites recorded with magic angle spinning. The resonances of Si atoms linked to n AlO$_4$ tetrahedra and $(4 - n)$ SiO$_4$ tetrahedra are labelled $n = 0-4$. The Si/Al ratios are as indicated. (Adapted from J. Klinowski, S. Ramdas, J. M. Thomas, C. A. Fyfe, and J. S. Hartman, *J. Chem. Soc. Faraday Trans. II*, **78** (1982) 1025.)

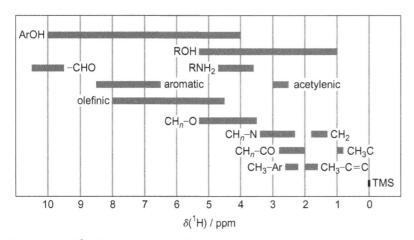

Fig. 2.12 Typical ^{1}H chemical shift ranges for some common organic functional groups relative to the reference compound tetramethylsilane ($\delta = 0$ ppm).

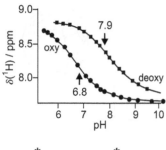

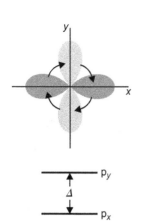

Fig. 2.13 The chemical shift of the H2 proton (asterisked) of histidine β146 in oxy- and deoxyhaemoglobin as a function of pH. (Adapted from I. D. Campbell and R. A. Dwek, *Biological spectroscopy*, Benjamin/Cummings, Menlo Park, CA, 1984, p. 161.)

Fig. 2.14 The circulation of electronic charge brought about by the mixing of electronic wave functions by a magnetic field. This paramagnetic current generates a small local magnetic field that deshields the nucleus at the centre of the electron density.

[HisH⁺]=[His], i.e. when the pH equals the pK_a of the imidazole group. Figure 2.13 shows the pH dependence of the chemical shift of H2 of one of the histidines (His β146) in the oxy and deoxy forms of haemoglobin, the protein responsible for oxygen transport in blood. The pK_a of His β146 in the deoxy form is higher by about 1 pH unit due to the stabilization of the HisH⁺ form by the CO_2^- group of a nearby aspartate (Asp β94). These two groups are brought into close proximity by the conformational changes in the protein that accompany deoxygenation. The interaction of His β146 and Asp β94 is partially responsible for the pH-dependence of the oxygen affinity of haemoglobin (the 'Bohr effect').

2.3 Origin of chemical shifts

A magnetic field can induce two kinds of electronic current in a molecule: *diamagnetic* and *paramagnetic* (a diamagnetic material is one in which the magnetization induced by an external field acts so as to *oppose* that field; in a paramagnetic material the induced magnetization *augments* the external field). Diamagnetic and paramagnetic currents flow in opposite directions and give rise to nuclear *shielding* and *deshielding*, respectively. The shielding constant may therefore be written as a sum of diamagnetic and paramagnetic contributions:

$$\sigma = \sigma_d + \sigma_p \tag{2.9}$$

with $\sigma_d > 0$ and $\sigma_p < 0$.

Diamagnetic currents arise from the circulation of electrons *within* atomic or molecular orbitals around the direction of the external field $\boldsymbol{B}_0$ (Fig. 2.2). The current so induced generates a small local field opposed to $\boldsymbol{B}_0$. The magnitude of the diamagnetic current is determined by the *ground state* electronic wave function of the atom or molecule, depends sensitively on the electron density close to the nucleus, and provides the only contribution to σ for spherical, closed-shell atoms. σ_d is fairly easy to calculate for atoms and varies strongly with the number of electrons: 17.8 ppm for hydrogen, 261 ppm for carbon, 961 ppm for phosphorus, rising to about 10,000 ppm for atoms in the fifth row of the periodic table.

Paramagnetic currents also arise from the movement of electrons within molecules, but by a more circuitous route. Imagine a somewhat artificial molecule with just two electronic states: a ground state comprising an atomic p_x orbital containing two electrons with paired spins, and an unoccupied, higher energy, excited state that resembles a p_y orbital (Fig. 2.14). An external magnetic field directed along the z-axis distorts the wave function of the ground state by mixing into it a small fraction of the excited state wave function. In this way, the field partially overcomes the energy gap between p_x and p_y, which would otherwise keep the electrons locked in p_x, and so creates a path for electrons to circulate in the *xy*-plane (Fig. 2.14). This induced current generates a magnetic field which (it turns out) *augments* the external field and *deshields* a nucleus at the centre of the electron density.

The extent of paramagnetic deshielding is clearly linked to the energy gaps involved: other things being equal, low-lying excited states should make a greater contribution than higher energy states. Theory suggests that σ_p should be approximately inversely proportional to Δ, the average excitation energy. The paramagnetic contribution σ_p is also related to the distance R between the nucleus and its surrounding electrons. Since the magnetic field at the centre of a small current loop is proportional to the inverse cube of its radius, we can expect a similar dependence for σ_p. So, very roughly,

$$\sigma_p \propto -\frac{1}{\Delta}\left\langle\frac{1}{R^3}\right\rangle \tag{2.10}$$

where $\langle\cdots\rangle$ indicates an average over the local electron distribution.

σ_p vanishes for a molecule with an axially symmetric local electron distribution (for instance, the π electrons in acetylene) when the external magnetic field is parallel to the symmetry axis. Similarly there is no paramagnetic contribution to the chemical shifts of atoms. The theory of diamagnetic and paramagnetic currents and the chemical shifts they generate is described by Atkins and Friedman (2011).

2.4 Contributions to nuclear shielding

It will be clear from the previous section that chemical shifts are sensitive to subtle changes in electronic structure and that calculating nuclear shielding constants from first principles is unlikely to be straightforward. For example, two isotopologues of carbon monoxide, $^{13}C^{16}O$ and $^{13}C^{17}O$, have ^{13}C chemical shifts that differ by 0.025 ppm. This small but measurable difference arises because the average bond length of $^{13}C^{16}O$ is 0.005% longer than that of $^{13}C^{17}O$ (as a result of the mass dependence of the vibrational frequency and the anharmonicity of the potential energy). Nevertheless, sophisticated *ab initio* electronic structure calculations, based on the fundamental assumptions of quantum mechanics, can now predict chemical shifts that are often in good agreement with experimental measurements. Such calculations are proving to be extremely useful for interpreting NMR spectra (e.g. Bonhomme et al. (2012)).

However, as we saw in Section 2.2, it is often not necessary to explain chemical shifts beyond the level of empirical correlations with structure (Fig. 2.12). Despite the complexities alluded to above, chemical shift differences *can* sometimes be traced back to straightforward changes in electron density or excited state energies. We explore these correlations in the following paragraphs.

It will prove useful to divide the nuclear shielding constant σ, arbitrarily, into four parts:

σ = local diamagnetic shielding
  + local paramagnetic shielding
  + shielding due to remote currents
  + other sources of shielding.   (2.11)

The first two are contributions from the electrons in the *immediate* vicinity of the nucleus, i.e. from electrons circulating around it. The third term accounts

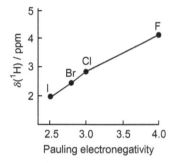

Fig. 2.15 ^{1}H chemical shifts of methyl halides plotted against the Pauling electronegativity of the halogen.

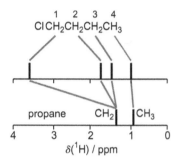

Fig. 2.16 Comparison of the ^{1}H chemical shifts of 1-chlorobutane and propane.

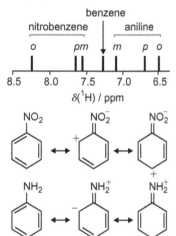

Fig. 2.17 Chemical shifts of the *ortho*, *meta* and *para* protons of aniline and nitrobenzene, together with resonance structures which account for the larger shielding/deshielding of the *ortho* and *para* protons compared to the *meta* proton.

for the diamagnetic and paramagnetic effects of electrons circulating around other (nearby) nuclei. The final part includes the effects of local electric fields, hydrogen bonds, solvent interactions, electron spins, etc. which, though diamagnetic or paramagnetic in origin, are more conveniently discussed separately. The remainder of this chapter gives a few illustrations of each of these contributions to σ.

Local diamagnetic shifts

The contributions to chemical shifts from local diamagnetic currents are strongly dependent on the electron density around the nucleus: the larger the electron density, the greater the shielding and the smaller the chemical shift, δ.

The ^{1}H chemical shifts of the methyl halides (Fig. 2.15) may readily be understood in these terms. As the electronegativity of the halogen increases, going from iodine to fluorine, electron density is withdrawn from the methyl group, deshielding the protons. Indeed there is a linear correlation between the ^{1}H chemical shifts and the Pauling electronegativities of the halogens. Methane, which lacks an electron withdrawing group, has a substantially smaller chemical shift (0.13 ppm) than methyl iodide. Electropositive substituents increase the shielding still further, e.g. 0.0 ppm (by definition) for tetramethylsilane.

The deshielding effect of electronegative atoms is fairly short range, as illustrated in Fig. 2.16. Relative to the CH_2 group in propane, the C1 protons of 1-chlorobutane are strongly deshielded; the effect on the C2 and C3 protons is much smaller, but still measurable. The terminal methyl group (C4), four carbons away from the chlorine, is essentially unaffected by the substituent, as judged by its chemical shift relative to the methyl protons in propane.

Similar behaviour is found for monosubstituted benzenes (Fig. 2.17). Groups with electron withdrawing mesomeric effects, NO_2 and CN for example, deshield the ring protons, while electron donating groups, such as NH_2 and OCH_3, result in shielding. The shielding/deshielding is most pronounced for the protons *ortho* and *para* to the substituent, an effect that can be rationalized by means of the resonance structures shown in the figure.

Local paramagnetic shifts

The Δ variation of σ_p (eqn 2.10) is nicely illustrated by the ^{59}Co chemical shifts of octahedral cobalt complexes. The five 3d orbitals of the cobalt(III) ion are split by the octahedral field of the ligands into a set of three degenerate t_{2g} orbitals and a pair of degenerate e_g orbitals (Fig. 2.18). The ground state corresponds to the $(t_{2g})^6$ electronic configuration. The four excited states that arise from the first excited configuration, $(t_{2g})^5(e_g)^1$, all have energies similar to the ligand field splitting, Δ. As shown in Fig. 2.18, there is a remarkably good correlation between the ^{59}Co chemical shift and the wavelength of the lowest energy absorption band which is roughly proportional to $1/\Delta$. Ligands such as carbonate, oxalate, and acetylacetonate which produce small ligand field splittings give large paramagnetic deshielding and hence large chemical shifts δ.

The ^{13}C chemical shifts of monosubstituted benzenes provide a good example of the dependence of σ_p on $\langle R^{-3} \rangle$ (eqn 2.10). As shown in Fig. 2.19, the *para* carbon is deshielded by electron withdrawing substituents (e.g. NO_2) and shielded by electron releasing groups (e.g. NH_2). Although this is the same trend observed for the ^1H shifts in these compounds, the effect is paramagnetic rather than diamagnetic in origin. Electron donating groups delocalize their lone pairs into the ring and increase the electron density at the *ortho* and *para* carbons. The increased electron repulsion causes the orbitals around these atoms to *expand*, reducing $\langle R^{-3} \rangle$ and hence δ.

At this point we can see why the chemical shift range of ^1H is so small compared to other nuclei ($\sim$10 ppm for ^1H; $\sim$200 ppm for ^{13}C; $\sim$300 ppm for ^{19}F, $\sim$500 ppm for ^{31}P, etc.). Local diamagnetic and local paramagnetic currents make only modest contributions because of the low electron density and high electronic excitation energy of the hydrogen atom. Indeed, ^1H chemical shifts are often more strongly influenced by the diamagnetic and paramagnetic currents in *neighbouring groups* of atoms which have larger electron densities and lower excitation energies, as we shall now see.

Neighbouring groups

To understand the (de)shielding of a nucleus caused by the motions of electrons in nearby groups of atoms, we take a specific example: an acetylene-substituted phenanthrene molecule (Fig. 2.20) in which the proton beside the C≡C bond is deshielded by 1.7 ppm compared to phenanthrene itself.

The magnetic field of the spectrometer ($\boldsymbol{B}_0$) induces electronic currents in the π electrons of the C≡C group which generate a local magnetic field. Let us suppose, for simplicity, that this small induced field has the same general form as the magnetic field that would be produced by a *magnetic dipole* $\boldsymbol{\mu}$ (i.e. a microscopic bar magnet) sitting at the centre of the C≡C bond. $\boldsymbol{\mu}$ is either parallel or antiparallel to $\boldsymbol{B}_0$ according to whether the induced current is paramagnetic or diamagnetic, respectively. Figure 2.21 shows two orientations of the phenanthrene molecule, with the acetylene group aligned perpendicular (a) and parallel (b) to $\boldsymbol{B}_0$. The dipolar field lines generated by $\boldsymbol{\mu}$ are drawn on the assumption that the diamagnetic currents are small (as is actually the case for acetylene). From the directions of the magnetic field lines, the phenanthrene proton should be deshielded in (a) and shielded in (b). Now we need to average over the random orientations of the molecule in solution which will inevitably cause (at least) partial cancellation of the shielding and deshielding effects shown in Fig. 2.21. In fact the cancellation would be *exact* if $\boldsymbol{\mu}$ were the same for all orientations of the molecule (Section 3.8). But the C≡C bond is magnetically anisotropic and currents are 'easier' to induce in some directions than in others. When the C≡C bond is parallel to $\boldsymbol{B}_0$ there is no paramagnetic current induced in the triple bond because of its cylindrical symmetry (Section 2.3). The diamagnetic π-electron currents are small for all orientations of the molecule so that the dominant effect occurs for orientation (a) in which the induced paramagnetic current is large. The result is that the phenanthrene proton is deshielded.

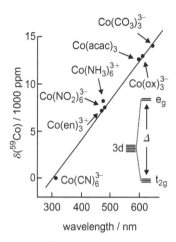

Fig. 2.18 ^{59}Co chemical shifts (relative to $Co(CN)_6^{3-}$) of octahedral cobalt complexes plotted against the wavelength of the first electronic absorption band. en = ethylenediamine; ox = oxalate; acac = acetylacetonate. (Adapted from R. Freeman, G. R. Murray, and R. E. Richards, *Proc. R. Soc. A*, **242** (1957) 455.)

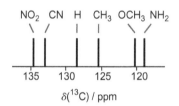

Fig. 2.19 ^{13}C chemical shifts of the *para* carbons in monosubstituted benzenes.

The magnetic field arising from a magnetic dipole is described in Appendix A.

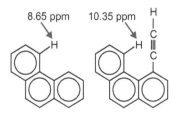

Fig. 2.20 Deshielding due to the magnetic anisotropy of a C≡C bond.

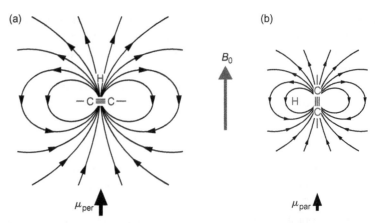

Fig. 2.21 Dipolar magnetic field lines (see Appendix A) generated by an induced magnetic moment at the centre of the acetylene group in the substituted phenanthrene shown in Fig. 2.20 (only the C≡C bond and the nearby phenanthrene proton are shown.). The proton is deshielded in orientation (a) and shielded in orientation (b) by the induced magnetic field. The opposite effect is expected for the acetylenic proton (not shown). The magnetic moment induced in the C≡C bond is larger in the perpendicular orientation (per) than the parallel orientation (par).

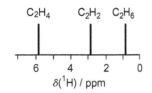

Fig. 2.22 ^{1}H chemical shifts of C_2H_6, C_2H_4, and C_2H_2.

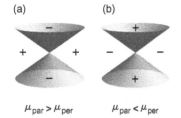

Fig. 2.23 Regions of shielding (+) and deshielding (−) due to neighbouring group magnetic anisotropy, for (a) $\mu_{par} > \mu_{per}$ and (b) $\mu_{par} < \mu_{per}$. If the fields generated by the induced magnetic moments were exactly those of a point dipole, the half angle of the cones would be $\theta = 54.7°$ (the so-called *magic angle* at which $3\cos^2\theta = 1$). The axis of the cones coincides with the symmetry axis of the neighbouring group.

This source of chemical shifts is generally referred to as *neighbouring group anisotropy*. The magnitude of the effect depends on the magnetic anisotropy of the neighbouring group itself and not on the nucleus being shielded or deshielded. It is therefore *relatively* more important for protons with their small local diamagnetic and paramagnetic currents than for other nuclei, ^{13}C for instance, which have larger local electron densities and lower excitation energies. Only groups that have very high symmetry, e.g. tetrahedral, have no magnetic anisotropy.

But what about the proton *attached* to the C≡C bond in acetylene? Figure 2.22 shows the ^{1}H chemical shifts of acetylene, ethylene, and ethane. Considering the hybridization of the molecular orbitals, one would expect $\delta(C_2H_6) < \delta(C_2H_4) < \delta(C_2H_2)$. As the s-electron character of the σ bonds increases (in the order $sp^3 < sp^2 < sp$) so the bonding electrons should be held more tightly to the carbons, removing electron density from the hydrogen atoms and deshielding the protons. The observed order of shifts arises from the *shielding* effect of the π electrons of the triple bond (Fig. 2.21(a) again), and also the neighbouring group effect of the C=C double bond which *deshields* the protons in ethylene.

A simple extension of the above argument leads to the diagrams in Fig. 2.23. Let μ_{par} and μ_{per} be the magnitudes of the induced magnetic moments when the $\boldsymbol{B}_0$ field is, respectively, parallel and perpendicular to the axis of the neighbouring group which is assumed to have cylindrical symmetry. When (a) $\mu_{par} > \mu_{per}$, a nucleus lying *within* one of the two cones should be deshielded, while a nucleus in the region *outside* the cones should be shielded. The opposite (b) holds when, as in acetylene, $\mu_{par} < \mu_{per}$. In reality, both the angular and radial dependence of the induced field are more complicated than this simple picture would suggest; nevertheless it remains a useful way of thinking about neighbouring group anisotropy, and if approached with caution, can be used to predict the direction, although not the magnitude, of chemical shift differences.

Another instance of the neighbouring group effect occurs in *aromatic* compounds, whose extensive π-electron 'clouds' can support large electronic currents. In molecules such as benzene, the dominant contribution to the magnetic anisotropy comes from the circulation of the π electrons *within* their delocalized molecular orbitals; i.e. the neighbouring group effect is due principally to the *diamagnetic* moment induced when the external field is parallel to the six-fold symmetry axis, as indicated in Fig. 2.24. However, the end result is the same as for acetylene because the induced diamagnetic moment is *opposed* to the external field ($\mu_{par} < 0$), so that the anisotropy ($\mu_{par} - \mu_{per}$) is once again negative. Thus we can anticipate deshielding for nuclei in the plane of the aromatic ring, and shielding for any nuclei above or below the ring.

This *ring current shift* is demonstrated clearly by benzene itself, whose ^{1}H chemical shift is +1.4 ppm relative to the olefinic protons in cyclohexa-1,3-diene (Fig. 2.25(a)). A more interesting example is the *trans*-dimethyl-substituted dihydropyrene in Fig. 2.25(b) which, with 14 π electrons, is aromatic according to the $4n + 2$ rule ($n = 3$). The ring protons are deshielded, as in benzene, but the methyl groups, which protrude above and below the plane of the molecule, are shielded by more than 5 ppm relative to ethane. (A negative δ value means that the resonance appears in the spectrum to the right of TMS.) As indicated in Fig. 2.24, the methyl protons lie in a region where the induced field *opposes* the external field.

Finally, the planar aromatic molecule [18]-annulene in Fig. 2.25(c), with 18 π electrons, shows two ^{1}H resonances, one from the 12 strongly deshielded external protons, and one at −2.99 ppm from the six internal protons. The latter set of hydrogen atoms lies within the current loop formed by the circulating π-electrons and, as indicated in Fig. 2.24, is shielded by the ring current effect. Note that the chemical shifts of these nuclei cannot, even qualitatively, be described using the point dipole approximation (Fig. 2.23) which predicts deshielding for any nucleus in the plane of the aromatic ring, however close to the centre of the molecule.

Although only C≡C and C=C bonds and aromatic rings have been discussed, neighbouring group effects exist for other functional groups, C–C, C=O, and N=O for example, and have a strong influence on ^{1}H chemical shifts.

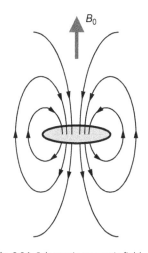

Fig. 2.24 Schematic magnetic field lines arising from the diamagnetic current induced in a benzene ring when the external field is parallel to the six-fold symmetry axis of the ring.

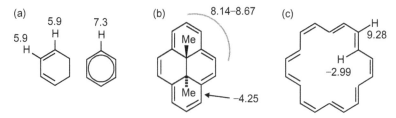

Fig. 2.25 Ring current shifts. ^{1}H chemical shifts (ppm) in: (a) benzene compared to cyclohexa-1,3-diene; (b) *trans*-15,16-dimethyl-15,16-dihydropyrene; (c) [18]-annulene.

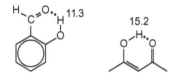

Fig. 2.26 Deshielding due to hydrogen bonding. ^{1}H chemical shifts (in ppm) in salicylaldehyde and the enol form of acetylacetone.

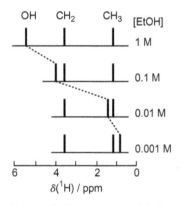

Fig. 2.27 Hydrogen bonding shifts in ethanol, as a function of concentration in CCl_4.

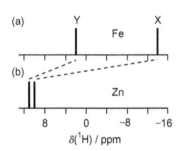

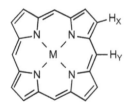

Fig. 2.28 ^{1}H chemical shifts of the two types of proton (X and Y) in (a) the paramagnetic Fe^{3+}(porphin) CN^- ion and (b) the diamagnetic Zn^{2+} (porphin) ion.

Other sources of chemical shifts

Hydrogen bonding is responsible for some of the largest observed ^{1}H chemical shifts. Two compounds that form intramolecular hydrogen bonds are shown in Fig. 2.26: in both, the hydrogen-bonded proton is heavily deshielded. Intermolecular hydrogen bonds, which are generally somewhat weaker, produce smaller shifts. For example, the hydroxyl proton resonance of ethanol moves by about −4 ppm when the intermolecular hydrogen bonding is disrupted by dilution in CCl_4 (see Fig. 2.27). The limiting shift of the OH resonance at infinite dilution (~0.8 ppm) is similar to the value for monomeric ethanol in the gas phase (0.55 ppm). Similar changes in the OH chemical shift are found with increasing temperature, which also favours the monomer side of the monomer ⇌ dimer, trimer, ... equilibrium. The origin of the deshielding caused by hydrogen bonding is unclear. Most likely, the strong electric field of the Y atom in X–H···Y draws the hydrogen atom slightly away from the electrons in the X–H bond, so reducing the electron density immediately around it.

Chemical shifts are also affected by the local *electric* fields arising from charged or polar groups. These can modify both diamagnetic and paramagnetic currents by polarizing local electron distributions, and by perturbing ground and excited state wave functions and energies. Positive charges usually deshield nearby protons, while negative charges often give rise to shielding. For example, the protons on the imidazole side-chain of the amino acid histidine are deshielded by about 1 ppm when the ring is protonated (Fig. 2.13).

Finally, in this far from complete survey, there are the *paramagnetic* shifts produced by electron spins (in this context paramagnetic refers to the permanent magnetic moment of the electron and not to the induced electronic currents discussed above). Unpaired electrons give rise to large dipolar magnetic fields (Appendix A)—the magnetogyric ratio of the electron is 660 times that of the proton—which can result in substantial nuclear shielding/deshielding, as illustrated by the two metal-porphin complexes in Fig. 2.28. The two types of proton in the aromatic ligand, which have large ring current shifts in the diamagnetic Zn^{2+} complex, are heavily shielded in the paramagnetic Fe^{3+} compound.

2.5 Summary

- Nuclei (e.g. protons, ^{1}H) in molecules have slightly different NMR frequencies, an effect known as the chemical shift.

- Chemical shifts arise from induced electronic currents which shield or deshield the nuclei from the applied magnetic field.

- The chemical shift parameter δ quantifies the extent of nuclear shielding/ deshielding.

- The magnitudes of induced diamagnetic and paramagnetic currents can be related to local electron densities and electronic excitation energies.

- Chemical shifts can often be understood by considering the effects of electron donating and withdrawing groups, induced currents in neighbouring groups, charged or polar groups, hydrogen bonds, and unpaired electrons.

- Chemical shifts distinguish nuclei in different environments in molecules and give information on molecular identity and structure.

2.6 Exercises

1. The ^{1}H chemical shifts of benzene and dimethyl sulphoxide are 7.3 and 2.4 ppm, respectively. What is the difference in the ^{1}H NMR frequencies of the two compounds on a 750 MHz spectrometer?

2. The ^{1}H Larmor frequency of C_2H_6 exceeds that of C_2H_4 by 3.0 kHz on a spectrometer with a 14.1 T magnet. The chemical shift of C_2H_6 is 0.9 ppm. What is the chemical shift of C_2H_4?

3. Use the data in Fig. 2.18 to estimate the difference in ^{59}Co NMR frequencies of $Co(CN)_6^{3-}$ and $Co(CO_3)_3^{3-}$ in a 9.4 T magnetic field. $[\gamma(^{59}Co) = 1.637 \times 10^6 \text{ T}^{-1} \text{ s}^{-1}]$.

4. The following compounds all exhibit a single line in their ^{1}H NMR spectra. Deduce their structures. (a) $C_6H_4Cl_2$. (b) $C_3H_6Cl_2$. (c) $C_3H_2Cl_6$. (d) $C_3H_4Cl_2$. (e) $C_6H_4O_2$.

5. (a) How many distinct chemical shifts would you expect to find in the ^{13}C spectra of the following isomers of C_5H_{12}: pentane, 2-methylbutane, and 2,2-dimethylpropane? (b) How many distinct chemical shifts would you expect to find in the ^{1}H spectra of the three isomers of dichlorocyclopropane?

6. ^{1}H and ^{13}C NMR spectra were recorded for two isomers of $C_3H_2Cl_6$. Both ^{13}C spectra contain peaks at three distinct chemical shifts. Isomer 1 has one distinct ^{1}H chemical shift and isomer 2 has two. (a) Deduce the structures of the two compounds. (b) Predict the number of chemical shifts in the ^{1}H and ^{13}C spectra of the other two isomers of $C_3H_2Cl_6$.

7. The lowest energy electronic transitions in alkanes and alkenes are approximately 10 eV and 8 eV, respectively. Predict whether saturated (sp^3) or unsaturated (sp^2) ^{13}C nuclei have larger chemical shifts.

8. Predict which of the following compounds has the highest and which the lowest ^{1}H chemical shift: CH_3Br, CH_2Br_2, $CHBr_3$.

9. Predict which of the *ortho*, *meta*, and *para* protons in methoxybenzene has the highest and which the lowest ^{1}H chemical shift.

10. The ring current shift of a ^{1}H nucleus close to a benzene ring is approximately proportional to $(1 - 3\cos^2 \theta)/r^3$, where θ is the angle between a vector perpendicular to the plane of the ring and the vector that connects the ^{1}H to the centre of the ring. r is the distance of the ^{1}H from the centre of the ring. The ring current shift of the protons in benzene is +2 ppm. The C–C and C–H bond lengths in benzene are 140 and 110 pm, respectively. A ^{1}H immediately above the centre of the ring ($\theta = 0$) has a ring current shift of –2 ppm. What is its distance from the centre of the ring?

Answers to the exercises are provided at the back of the book. Full worked solutions are available on the Online Resource Centre at www.oxfordtextbooks.co.uk/orc/hore_nmr2e/

3 Spin–spin coupling

3.1 Introduction

Chapter 2 may have given the impression that the appearance of liquid-state NMR spectra is determined solely by chemical shifts—one resonance for each distinct nuclear environment. In fact, there is another extremely valuable source of information encoded in most NMR spectra, namely the magnetic interactions between nuclei, known variously as *spin–spin couplings*, *scalar couplings*, or *J-couplings*. Amongst other things, these interactions cause the ^{1}H spectrum of liquid ethanol to comprise not three (Fig. 1.1) but eight (and sometimes more) resonances when recorded at high resolution (Fig. 3.1).

3.2 Effect on NMR spectra

As Fig. 3.1 suggests, nuclear spin–spin coupling causes NMR lines to split into a small number of components with characteristic relative intensities and spacings. In the case of ethanol, the CH$_3$ peak becomes a *triplet*—three equally spaced

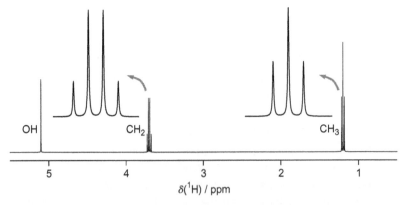

Fig. 3.1 400 MHz ^{1}H NMR spectrum of liquid ethanol showing the fine structure produced by spin–spin coupling. Compare this spectrum with Fig. 1.1, in which the splittings are obscured by instrumental line broadening. Further structure appears in the spectrum when all traces of acid or base are removed (Fig. 4.12).

lines with relative amplitudes in the ratio 1:2:1—and the CH_2 resonance is split into a *quartet*—four equally spaced lines with relative intensities 1:3:3:1. To see how this *multiplet* structure arises we focus initially on a much simpler molecule, the formate ion HCO_2^- in which the carbon is ^{13}C.

On the basis of the previous chapter, we might expect to see a single NMR line in each of the ^1H and ^{13}C spectra. In fact both spins give rise to 'doublets': two lines disposed symmetrically either side of the chemical shift position, as shown in Fig. 3.2. The splitting (195 Hz in this case) is the strength of the ^1H–^{13}C spin–spin interaction and is the same for both spectra.

The ^1H resonance is split into two because the magnetic moment of the ^{13}C produces a small local magnetic field at the position of the ^1H. When the ^{13}C is in its $m = +\frac{1}{2}$ state (here denoted C↑), it generates a magnetic field that opposes the external field and shifts the ^1H resonance to the right in Fig. 3.3. Conversely, for an $m = -\frac{1}{2}$ carbon (C↓), the local field adds to the external field and moves the ^1H resonance in the opposite direction. In the language of chemical shifts, a C↑ carbon shields the ^1H and a C↓ carbon deshields it. The two components of the ^1H doublet thus correspond to two sorts of $H^{13}CO_2^-$ molecule: those with C↑ and those with C↓. Since the difference in energy between the two configurations of the ^{13}C spin is tiny compared to k_BT, the two kinds of $H^{13}CO_2^-$ are equally likely, and the two components of the ^1H doublet are equally intense. An exactly analogous argument explains the splitting of the ^{13}C resonance by the ^1H.

It is evident from Fig. 3.3 that the spin–spin interaction in $H^{13}CO_2^-$ stabilizes the antiparallel arrangements of nuclear spins (H↑C↓ and H↓C↑) and destabilizes the parallel configurations (H↑C↑ and H↓C↓). Thus the two energy levels of the proton ($m = \pm\frac{1}{2}$) are each split into two, with energies determined by the relative orientations of the ^{13}C and ^1H spins. The ^1H NMR transitions (H↑C↑ → H↓C↑) of molecules containing a C↑ have a lower energy because the transition is from an energetically unfavourable state (parallel spins) to a favourable one (antiparallel spins). Conversely, molecules containing C↓ have higher energy ^1H transitions (H↑C↓ → H↓C↓).

A *heteronuclear* example (^1H–^{13}C) has been used to illustrate the nature of J-coupling merely as a matter of convenience; *homonuclear* couplings, e.g. between two protons with different chemical shifts, give rise to splittings in exactly the same way.

The properties of spin–spin coupling as illustrated by $H^{13}CO_2^-$ may be summarized and generalized in the following simple expression for the energy levels of two interacting nuclei A and X (not necessarily spin-$\frac{1}{2}$):

$$E(m_A, m_X) = m_A h\nu_{0A} + m_X h\nu_{0X} + hJ_{AX}m_A m_X \tag{3.1}$$

in which m_A and m_X are the magnetic quantum numbers of the two nuclei and ν_{0A} and ν_{0X} are the Larmor frequencies (eqn 2.4). J_{AX} is the strength of the interaction, known as the *spin–spin coupling constant* or the *J-coupling constant*. It is measured in frequency units (hertz) and may be positive or negative: if the antiparallel arrangement of nuclear spins is energetically favoured, then $J_{AX} > 0$ (as in $H^{13}CO_2^-$); when the parallel spin configuration is lower in energy, $J_{AX} < 0$.

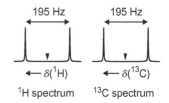

195 Hz   195 Hz

← $\delta(^1H)$ ← $\delta(^{13}C)$

^1H spectrum ^{13}C spectrum

Fig. 3.2 ^1H and ^{13}C spectra of $H^{13}CO_2^-$ showing the doublets produced by ^1H–^{13}C J-coupling. The arrowheads indicate the chemical shift positions at the centre of each doublet.

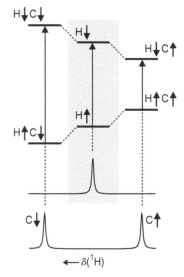

Fig. 3.3 The effect of ^1H–^{13}C J-coupling in $H^{13}CO_2^-$ on the energy levels and spectrum of the ^1H spin. For clarity, the energy-level shifts due to the J-coupling have been greatly exaggerated. The central pair of energy levels and the upper spectrum are appropriate in the *absence* of a spin–spin interaction. J-coupling produces the energy levels on the left and right, and the lower spectrum.

Equation 3.1 is valid when $|\nu_{0A} - \nu_{0X}| \gg |J_{AX}|$. See Section 3.5.

Combining eqn 3.1 with the selection rule $\Delta m_A = \pm 1$, one can see that the A–X interaction shifts the Larmor frequency of spin A (v_{0A}, eqn 2.4) by $J_{AX}m_X$. More generally, the equation for the resonance frequency becomes:

$$v_A = v_{0A} + \sum_{K \neq A} J_{AK}m_K, \qquad (3.2)$$

where the summation runs over all spins (K) that have a non-negligible J-coupling with A.

Figure 3.4 shows the complete energy-level diagram and the corresponding NMR spectra for a pair of spin-$\frac{1}{2}$ nuclei with and without J-coupling. Note that the allowed transitions are those in which just one spin changes its magnetic quantum number ($\Delta m_A = \pm 1$ or $\Delta m_X = \pm 1$). Simultaneous changes in m_A and m_X, i.e. A↑X↑ ↔ A↓X↓ and A↑X↓ ↔ A↓X↑, are forbidden.

It should be clear from eqns 3.1 and 3.2 that the *sign* of the coupling constant has no effect on the appearance of the spectrum. For example, changing J_{AX} in

Remember that $v_{0A} < 0$ for a nucleus with $\gamma_A > 0$, so that when $J_{AK}m_K > 0$, the resonance is shifted in the direction of increasing, i.e. less negative, frequency and thus moves to the right in the spectrum.

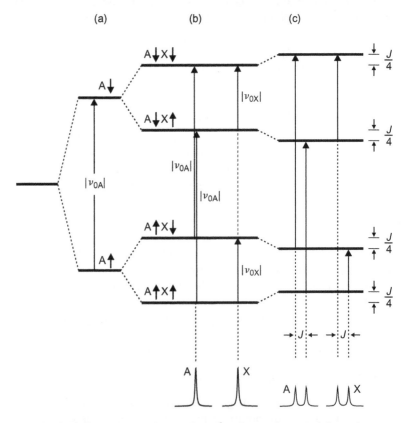

Fig. 3.4 Energy levels and spectra of a pair of spin-$\frac{1}{2}$ nuclei, A and X. From left to right, magnetic interactions are introduced in the order: (a) the interaction of A with the magnetic field B_0; (b) the interaction of X with B_0; (c) the spin–spin coupling, $J = J_{AX}$. For clarity, the energy-level shifts are not drawn to scale. v_{0A} and v_{0X} are the Larmor frequencies of the two spins in the absence of coupling. The shifts in the energy levels are given as frequencies. The figure is drawn for $\gamma_A > 0$, $\gamma_X > 0$ (so that v_{0A} and v_{0X} are both negative) and $J_{AX} > 0$.

Fig. 3.4 from positive to negative simply interchanges the two components of each doublet.

These simple ideas, exemplified by $H^{13}CO_2^-$, and embodied in eqns 3.1 and 3.2, allow one to predict the effect of spin–spin coupling on the NMR spectrum of almost any molecule. The exceptions will be dealt with later (Section 3.5).

3.3 Multiplet patterns

Having seen that coupling between nuclear spins can affect NMR spectra, we now look at some frequently encountered spin systems (collections of coupled nuclei) to see how they give rise to distinctive multiplets (doublets, triplets, quartets, etc.).

At this stage it is assumed that all pairs of spins are *weakly coupled,* i.e. that the difference in Larmor frequencies of the two nuclei, $|v_{0A} - v_{0X}|$, greatly exceeds their mutual coupling, $|J_{AX}|$. The complications associated with *strong coupling* are discussed in Section 3.5. All nuclei are spin-$\frac{1}{2}$, unless otherwise stated. The term *equivalent nuclei* is used to describe spins in identical environments, with identical chemical shifts—for example the protons in CH_4 or the fluorines in CF_3COOH. This somewhat loose definition will be refined at the end of this section.

The following paragraphs deal with the effect of spins M and X on the NMR signal of spin A. The convention is that spins with very different chemical shifts are labelled by letters far apart in the alphabet (e.g. A, M, X). Nuclei having similar shifts, and thus likely to be strongly coupled, are assigned adjacent letters in the alphabet (e.g. A, B, C).

Finally, it must be said that the predictions in the following paragraphs are not infallible: the expected multiplet patterns may be obscured if the splitting is smaller than the linewidth (see Chapter 5), or modified if the molecule is undergoing a dynamic process that causes the J-couplings to be time-dependent (see Chapter 4).

Coupling to a single spin-$\frac{1}{2}$ nucleus (AX)

As already discussed for $H^{13}CO_2^-$, the interaction of nucleus A with a single spin-$\frac{1}{2}$ nucleus, X, causes the A resonance to split into two equally intense lines centred at the chemical shift of A (a doublet), with spacing equal to the AX coupling constant, J_{AX}. The interaction is symmetrical, so that the spectrum of X is also a doublet, with the same splitting (Figs 3.2–3.4).

Coupling to two inequivalent spin-$\frac{1}{2}$ nuclei (AMX)

The next level of complexity is the AMX spin system, which consists of three nuclei with different chemical shifts and three distinct coupling constants: J_{AM}, J_{AX}, J_{MX}. Equation 3.2 can be used to predict the spectrum of A, by drawing up a list of the possible values of the magnetic quantum numbers of M and X (Table 3.1). Four lines are expected because there are four non-degenerate arrangements of the

Table 3.1 Spin–spin coupling in an AMX spin system

m_M	m_X	$\sum_{K=M,X} J_{AK}m_K$
$+\frac{1}{2}$	$+\frac{1}{2}$	$+\frac{1}{2}(J_{AM}+J_{AX})$
$+\frac{1}{2}$	$-\frac{1}{2}$	$+\frac{1}{2}(J_{AM}-J_{AX})$
$-\frac{1}{2}$	$+\frac{1}{2}$	$-\frac{1}{2}(J_{AM}-J_{AX})$
$-\frac{1}{2}$	$-\frac{1}{2}$	$-\frac{1}{2}(J_{AM}+J_{AX})$

The final column gives the shift in the Larmor frequency of A for each of the four spin configurations of M and X (both $I=\frac{1}{2}$) (see eqn 3.2).

M and X spins (M↑X↑, M↑X↓, M↓X↑, M↓X↓). These peaks are displaced from the chemical shift of A by simple combinations of J_{AM} and J_{AX} (but not J_{MX}). The A multiplet should therefore be a *doublet of doublets*, as shown in Fig. 3.5(a).

A different way to see how this pattern arises is to construct the spectrum in stages using the 'tree diagram' approach shown in Fig. 3.5(a). Imagine first of all that both J_{AM} and J_{AX} are zero, so that the spectrum of A is a singlet at the chemical shift position. Now suppose the AM coupling is 'switched on', to give a doublet with splitting J_{AM}. Finally, when the AX coupling is introduced, each of the lines of the doublet is itself split into a doublet, with splitting J_{AX}. This stepwise procedure is probably the quickest way of arriving at multiplet patterns. The order in which the couplings are introduced is irrelevant. Of course, the exact appearance of the doublet of doublets will depend on the values (but not the signs) of the coupling constants. This point is illustrated later (Fig. 3.13) for a *four*-spin system.

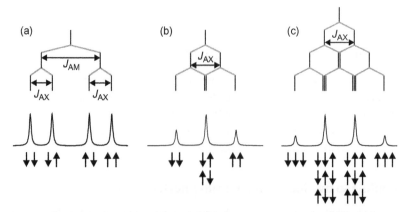

Fig. 3.5 (a) The NMR spectrum of nucleus A in an AMX spin system. The four components of the A multiplet, a doublet of doublets, arise from the four combinations of M and X magnetic quantum numbers, indicated ↑ $(m=+\frac{1}{2})$ and ↓ $(m=-\frac{1}{2})$. (b) The spectrum of nucleus A in an AX_2 spin system. (c) The spectrum of nucleus A in an AX_3 spin system. The spectra are drawn for $J_{AM}>J_{AX}>0$. The tree diagrams above the spectra show how the multiplet patterns arise.

Coupling to two equivalent spin-$\frac{1}{2}$ nuclei (AX$_2$)

This is a special case of the AMX spin system, with $J_{AM}=J_{AX}$. As may be seen from Table 3.1 and Fig. 3.5(b), the two central lines of the doublet of doublets coincide so that the multiplet becomes a *triplet* centred at the chemical shift of A, with line-spacing equal to the coupling constant, and relative intensities 1:2:1. The central line of the triplet arises from two degenerate arrangements of the X spins ($\uparrow\downarrow$ and $\downarrow\uparrow$), in both of which the local magnetic fields due to the X nuclei exactly cancel.

Coupling to three equivalent spin-$\frac{1}{2}$ nuclei (AX$_3$)

The multiplet pattern of A in an AX$_3$ spin system (three identical AX coupling constants) is a four-line quartet (Fig. 3.5(c) and Table 3.2). There are two peaks displaced from the chemical shift position by $\pm\frac{3}{2}J_{AX}$ and two peaks with three times the intensity at $\pm\frac{1}{2}J_{AX}$. The inner lines, for example, have relative intensity 3 because there are three degenerate ways of achieving a total magnetic quantum number of $\pm\frac{1}{2}$.

Coupling to n equivalent spin-$\frac{1}{2}$ nuclei (AX$_n$)

It should be clear how the results for AX, AX$_2$, and AX$_3$ can be generalized. For n equivalent X nuclei, the A resonance is split into $n+1$ equally spaced lines, with

Table 3.2 Spin–spin coupling in an AX$_3$ spin system

m_1	m_2	m_3	$\sum\limits_{i=1,2,3} J_{AX}m_i$
$+\frac{1}{2}$	$+\frac{1}{2}$	$+\frac{1}{2}$	$+\frac{3}{2}J_{AX}$
$+\frac{1}{2}$	$+\frac{1}{2}$	$-\frac{1}{2}$	
$+\frac{1}{2}$	$-\frac{1}{2}$	$+\frac{1}{2}$	$+\frac{1}{2}J_{AX}$
$-\frac{1}{2}$	$+\frac{1}{2}$	$+\frac{1}{2}$	
$+\frac{1}{2}$	$-\frac{1}{2}$	$-\frac{1}{2}$	
$-\frac{1}{2}$	$+\frac{1}{2}$	$-\frac{1}{2}$	$-\frac{1}{2}J_{AX}$
$-\frac{1}{2}$	$-\frac{1}{2}$	$+\frac{1}{2}$	
$-\frac{1}{2}$	$-\frac{1}{2}$	$-\frac{1}{2}$	$-\frac{3}{2}J_{AX}$

The final column shows the shift in the Larmor frequency of A for each of the eight spin configurations of the three X spins ($I=\frac{1}{2}$), labelled 1, 2, and 3 (see eqn 3.2).

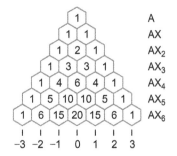

Fig. 3.6 Pascal's triangle showing the binomial coefficients in the expansion of $(1 + x)^n$. The rows give the relative intensities of the $n + 1$ lines in the A multiplet of an AX_n spin system ($n = 0$–6), where X is a spin-$\frac{1}{2}$ nucleus. As indicated at the bottom of the figure, the columns give the positions of the lines relative to the chemical shift position, in units of J_{AX}.

relative intensities given by simple combinatorial arithmetic. The amplitude of the m-th line ($m = 0, 1, 2, \cdots n$) of an AX_n multiplet is simply the number of ways in which m spins can be $\uparrow$ and $(n - m)$ spins $\downarrow$, i.e. $n!/m!(n - m)!$. To put it another way, the amplitudes are given by the coefficients in the binomial expansion of $(1 + x)^n$, or, equivalently, by the $(n + 1)$-th row of Pascal's triangle (Fig. 3.6).

Coupling involving $I > \frac{1}{2}$ nuclei

If the nucleus of interest, A, has spin quantum number greater than $\frac{1}{2}$, its multiplet structure can be predicted in exactly the same way as for a spin-$\frac{1}{2}$ nucleus. This can be seen from eqns 3.1 and 3.2 and is demonstrated in Fig. 3.7 for a spin-1 coupled to a spin-$\frac{1}{2}$. For example, the ^{14}N ($I = 1$) and ^{15}N $\left(I = \frac{1}{2}\right)$ NMR spectra of, respectively, $^{14}NH_4^+$ and $^{15}NH_4^+$ both consist of a quintet, with relative peak

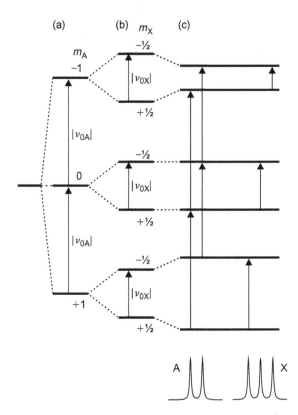

Fig. 3.7 Energy levels and spectra of a spin-1 nucleus (A) coupled to a spin-$\frac{1}{2}$ nucleus (X). From left to right, magnetic interactions are introduced in the order: (a) the interaction of A with the magnetic field B_0; (b) the interaction of X with B_0; (c) the spin-spin coupling, $J = J_{AX}$. The spectrum of A is a doublet because its four allowed NMR transitions are pairwise degenerate. The spectrum of X comprises three lines arising from the $m = +1, 0, -1$ states of A. For clarity, the energy-level shifts are not drawn to scale. ν_{0A} and ν_{0X} are the Larmor frequencies of the two spins in the absence of coupling. The shifts in the energy levels are given as frequencies. The figure is drawn for $\gamma_A > 0$, $\gamma_X > 0$ (so that ν_{0A} and ν_{0X} are both negative) and $J_{AX} < 0$. The energy levels are labelled with the appropriate magnetic quantum numbers.

intensities 1:4:6:4:1. The NH coupling constants of the two isotopologues are in the ratio 0.713:1, which is the ratio of the magnetogyric ratios of the two nitrogen isotopes (see Table 1.3).

However, nuclei with $I > \frac{1}{2}$ possess, in addition to their magnetic dipole moment, an *electric quadrupole moment* that can interact with local *electric field gradients*. For molecules tumbling in solution, this interaction can lead to efficient relaxation of the quadrupolar nucleus giving NMR lines that may be so broad that the expected multiplet patterns are partially or completely obscured. This quadrupolar relaxation mechanism is discussed further in Section 5.7.

The term 'tumbling' denotes the rapid chaotic rotational motion of a molecule in a liquid. Collisions with other molecules cause frequent changes in the axis and rate of rotation.

For A ($I = \frac{1}{2}$) coupled to X ($I > \frac{1}{2}$), the principles established above for spin-$\frac{1}{2}$ nuclei can easily be extended. A spin-I particle has energy levels corresponding to $2I + 1$ orientations of its magnetic moment with respect to the magnetic field B_0. Therefore, a nucleus coupled to a single X spin with quantum number I should show a multiplet comprising $2I + 1$ lines with equal spacings and amplitudes. For example, the ^{13}C spectrum of deuterated chloroform, $^{13}CDCl_3$, is a 1:1:1 triplet arising from the three equally probable states of the deuteron, $m = +1, 0, -1$ (Fig. 3.7). Once again, quadrupolar relaxation may upset these predictions. Rapid relaxation of the quadrupolar nucleus may have the effect of 'decoupling' A and X, such that no splitting is observed in the spectrum of A. For example, ^{35}Cl and ^{37}Cl (both $I = \frac{3}{2}$) rarely produce splittings in the NMR spectra of nearby nuclei. We shall return to this point in Section 5.7.

For coupling to *equivalent* $I > \frac{1}{2}$ nuclei, the multiplet patterns are easily deduced using the 'tree diagram' approach introduced in Fig. 3.5. For instance, the terminal protons of $^{11}B_2H_6$ (diborane) show a 1:1:1:1 quartet due to coupling to the directly bonded ^{11}B ($I = \frac{3}{2}$), while the bridge protons exhibit a seven-line pattern with relative intensities 1:2:3:4:3:2:1, arising from equal interactions with the two symmetrically placed borons (Fig. 3.8).

Equivalent nuclei

Up to now, we have used the term *equivalent* somewhat loosely to describe nuclei with identical chemical shifts, usually as a result of molecular symmetry. In fact there are two kinds of equivalence: *chemical* and *magnetic*. The distinction is

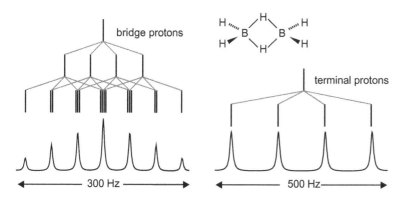

Fig. 3.8 ^1H NMR spectra of the terminal and bridge protons in diborane, $^{11}B_2H_6$.

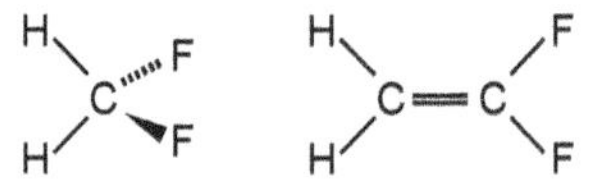

Fig. 3.9 CH_2F_2 (magnetically equivalent protons) and $CH_2=CF_2$ (chemically equivalent protons).

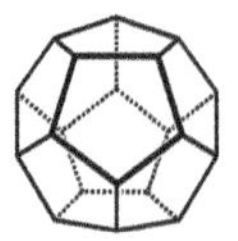

$\delta(^1H) = 3.38$ ppm

Fig. 3.10 Dodecahedrane, $^{12}C_{20}H_{20}$. The 20 protons are magnetically equivalent.

best seen by means of an example. Consider the protons in the two compounds CH_2F_2 and $CH_2=CF_2$ (Fig. 3.9). In CH_2F_2, the two protons have the same chemical shift *and* each has identical couplings to each of the fluorines: as such they are termed *magnetically equivalent*. The same cannot be said of $CH_2=CF_2$, where the *cis* and *trans* $^1H-^{19}F$ coupling constants differ: in this case the protons are said to be *chemically equivalent*.

More generally, a set of nuclei (*a*, *b*, *c*, . . .) with identical chemical shifts are magnetically equivalent *either* if there are no other spins in the molecule *or* if, for every other nucleus (e.g. *z*) in the molecule, the spin–spin coupling constants satisfy the relation

$$J_{az} = J_{bz} = J_{cz} = \cdots .$$
(3.3)

As might be expected, the NMR spectra of molecules containing chemically equivalent spins are rather more complex than for similar compounds with magnetically equivalent nuclei. For example, the ^1H spectrum of $CH_2=CF_2$ has no fewer than ten lines. The analysis of such spectra is not straightforward and will not be attempted here: a good discussion is given by Günther (2013). In the remainder of this section we concentrate on magnetically equivalent spins.

The ^1H spectrum of CH_2F_2 comprises just three lines: a 1:2:1 triplet with splitting equal to the proton-fluorine coupling constant J_{HF} (^{19}F is spin-$\frac{1}{2}$). The remarkable thing about this spectrum is not the triplet, which is exactly what one would expect for a *single* proton coupled to two identical fluorines, but the *absence* of any splittings arising from the $^1H-^1H$ coupling. Although the two protons interact (they are only two bonds apart), their mutual coupling is not manifest as a splitting in the spectrum. This is a general feature of *J*-coupling: *spin–spin interactions within a group of magnetically equivalent nuclei do not produce multiplet splittings.*

Perhaps without realizing it, we have already seen several instances of this phenomenon: each of the five molecules in Fig. 2.6 contains a single group of (magnetically) equivalent protons and each gives rise to an NMR singlet. A more esoteric example is the highly symmetrical molecule dodecahedrane (Fig. 3.10) whose ^1H spectrum also consists of a single peak.

The high resolution spectrum of ethanol in Fig. 3.1 can now be understood. The ethyl protons make up an A_3X_2 spin system: the triplet arises because each of the CH_3 protons couples equally to the two equivalent CH_2 protons, while the quartet comes from the CH_2 protons interacting identically with each of the CH_3 protons. As discussed in Chapter 4, rapid internal rotation around the C–C bond averages out the chemical shift differences associated with the different conformations of the molecule, and effectively renders the three methyl protons magnetically equivalent to one another, and similarly the two methylene protons. The absence of splittings from coupling between the CH_2 group and the OH proton is another story, also told in Chapter 4.

In Section 3.5, we shall see *why* magnetically equivalent nuclei do not split one another's NMR lines, but first a few examples that illustrate how multiplet patterns can be used to determine or verify the structures of molecules without prior knowledge of the magnitudes of the chemical shifts or coupling constants involved.

3.4 **Examples**

Figure 3.11 shows the very different ^{31}P NMR spectra of three closely related phosphorus–sulphur compounds: α-P$_4$S$_4$, β-P$_4$S$_4$, and β-P$_4$S$_5$. The multiplet structure arises entirely from ^{31}P–^{31}P couplings because ^{32}S, the only isotope of sulphur with an appreciable natural abundance (99.24%), has spin $I=0$. The three spin systems A$_4$ (α-P$_4$S$_4$), AMX$_2$ (β-P$_4$S$_4$), and A$_2$X$_2$ (β-P$_4$S$_5$) are easily deduced from the spectra, and are clearly consistent with the structures shown.

The tetrameric structure of t-butyllithium is clearly revealed by low temperature ^{13}C and ^{7}Li NMR (Fig. 3.12). The ^{7}Li spectrum of ^{7}Li^{13}CMe$_3$ consists of a 1:3:3:1 quartet: each lithium interacts with three equivalent t-butyl carbons, and has an unresolved (i.e. very small) coupling to the fourth, more distant ^{13}C. Similarly, the ^{13}C spectrum of ^{6}Li^{13}CMe$_3$ is a septet, with relative intensities 1:3:6:7:6:3:1, produced by each of the four equivalent ^{13}C spins interacting with three equivalent $I=1$ ^{6}Li nuclei. The two coupling constants, $J(^7$Li^{13}C$)=14.3$ Hz and $J(^6$Li^{13}C$)=5.4$ Hz, are in the ratio of the magnetogyric ratios of the two Li isotopes (1.04×10^8 and 3.94×10^7 T^{-1} s^{-1}, respectively). As discussed in Chapter 4, these spectra are modified at higher temperatures by rapid rearrangement of the t-butyl groups.

A slightly more complex case is the ^{1}H spectrum of 1,3-bromonitrobenzene (Fig. 3.13). This is a weakly coupled AMPX spin system with all six pairwise couplings resolved, so that each proton gives a doublet of doublets of doublets, i.e. eight lines. The exact appearance of each multiplet is determined by the magnitudes of the coupling constants, and may readily be understood by noting that $|J_{ortho}|>|J_{metal}|>|J_{para}|$. For the A and X multiplets, the central pair of lines overlap strongly and appear as a single line of double intensity. Two further illustrations of the use of spin–spin couplings in structural studies are given in Chapter 6 (Figs 6.18 and 6.19).

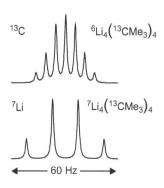

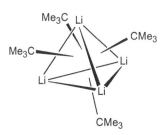

Fig. 3.12 ^{13}C spectrum of ^{6}Li$_4$(^{13}CMe$_3$)$_4$ and ^{7}Li spectrum of ^{7}Li$_4$(^{13}CMe$_3$)$_4$. Both spectra were recorded using ^{1}H-decoupling to remove the multiplet splittings caused by the ^{1}H nuclei.

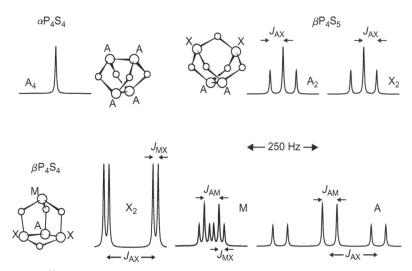

Fig. 3.11 ^{31}P NMR multiplets of α-P$_4$S$_4$, β-P$_4$S$_4$, and β-P$_4$S$_5$. The larger spheres represent the phosphorus atoms.

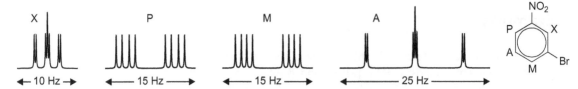

Fig. 3.13 ^{1}H NMR spectrum of 1,3-bromonitrobenzene. The six coupling constants are: $J_{AM} = 7.98$ Hz; $J_{AP} = 8.28$ Hz; $J_{MP} = 0.99$ Hz; $J_{MX} = 1.89$ Hz, $J_{PX} = 2.18$ Hz, $J_{AX} = 0.34$ Hz.

^{13}C NMR

As an NMR nucleus, ^{13}C is second in popularity only to ^{1}H; it is therefore appropriate at this point to comment briefly on multiplet splittings in ^{13}C spectra. In organic molecules, the dominant couplings experienced by ^{13}C nuclei are with their directly bonded protons. One-bond C–H coupling constants generally fall in the range 100–250 Hz, and are often an order of magnitude larger than two-bond and three-bond C–H interactions. The ^{13}C multiplets produced by one-bond couplings—a quartet for a methyl carbon (CH_3), a triplet for a methylene (CH_2), a doublet for a methine (CH), and a singlet for a quaternary carbon (C)—provide valuable clues when attempting to assign peaks in a spectrum to particular carbons in the molecule. However, ^{13}C NMR spectra are normally measured with the protons *decoupled*, so as to remove the ^{13}C–^{1}H splittings. This is achieved by irradiating the sample at the ^{1}H resonance frequency (about four times that of ^{13}C) while the ^{13}C spectrum is being recorded. The result is a considerably simplified spectrum: in the absence of heteronuclei (^{19}F, ^{31}P, etc.) each inequivalent carbon site in a molecule gives rise to a singlet in the ^{1}H-decoupled ^{13}C spectrum (denoted ^{13}C{^{1}H}).

Not only are ^{13}C{^{1}H} spectra less crowded than those with the proton–carbon couplings present, they also have higher sensitivity. The latter arises from the nuclear Overhauser enhancement (a relaxation phenomenon described in Section 5.5) *and* because all the NMR intensity for each multiplet is concentrated into a single line.

Finally, *homonuclear* (^{13}C–^{13}C) couplings are not normally observed in ^{13}C spectra because of the low natural abundance of ^{13}C (1.1%). Taking ethanol as an example, it is clear that of the molecules containing a ^{13}C at a given position, only about 1 in 100 contains a second ^{13}C. Thus, the spectrum of ^{13}CH$_3$^{13}CH$_2$OH should be about 100 times weaker than that of either ^{12}CH$_3$^{13}CH$_2$OH or ^{13}CH$_3$^{12}CH$_2$OH. ^{13}C–^{13}C splittings therefore often go unnoticed. Of course ^{12}CH$_3$^{12}CH$_2$OH, by far the most abundant isotopologue, has no ^{13}C NMR spectrum at all. For more on ^{13}C NMR see Wehrli et al. (1988), Friebolin (2011), and Günther (2013).

3.5 Strong coupling and equivalent spins

In Sections 3.2–3.4 we saw that the spectrum of a pair of coupled spin-$\frac{1}{2}$ nuclei can either be two doublets (weak coupling) or one singlet (magnetic equivalence). To shed some light on this, and on what happens between these two

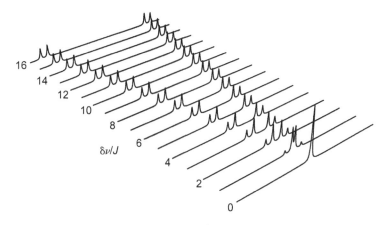

Fig. 3.14 Calculated NMR spectra of a pair of spin-$\frac{1}{2}$ nuclei for fixed J and a range of values of δv.

extremes, we start with Fig. 3.14 which shows spectra calculated for a range of values of $\delta v = v_{0A} - v_{0B}$ (the difference in chemical shift frequencies of spins A and B). Keeping the J-coupling fixed, the two doublets move together as their chemical shifts become more similar. At the same time, the inner components of the four-line pattern steadily become stronger while the outer components become weaker. Eventually, when $\delta v = 0$, the inner lines coincide and the outer lines vanish.

We denote the $m = +\frac{1}{2}$ and $m = -\frac{1}{2}$ states of each spin α and β, respectively, so that the four states of the weakly coupled pair are

$$\alpha_A \alpha_B, \quad \alpha_A \beta_B, \quad \beta_A \alpha_B, \quad \beta_A \beta_B.$$

As described in Section 3.2, the four allowed transitions,

$$\alpha_A \alpha_B \leftrightarrow \alpha_A \beta_B, \quad \alpha_A \alpha_B \leftrightarrow \beta_A \alpha_B, \quad \alpha_A \beta_B \leftrightarrow \beta_A \beta_B, \quad \beta_A \alpha_B \leftrightarrow \beta_A \beta_B,$$

have distinct frequencies and equal intensities giving the familiar pair of doublets (Fig. 3.15(a)). As δv becomes smaller, the separation of the two central states becomes comparable to J with the result that they *mix*. Instead of being pure $\alpha_A \beta_B$ and $\beta_A \alpha_B$ they become *linear combinations* of $\alpha_A \beta_B$ and $\beta_A \alpha_B$. The consequence is a change in the *transition probabilities* and *transition frequencies*. The inner lines become more allowed (i.e. stronger) and the outer pair less allowed (weaker), the effect being more pronounced as $\delta v/J$ becomes smaller (Fig. 3.15(b)). In the limit $\delta v = 0$, the outer lines are forbidden and the inner lines have the same frequency (Fig. 3.15(c)).

To make this more concrete, we now summarize the results of a quantum mechanical treatment (Hore, Jones, and Wimperis (2015)). The four states ψ_j and their energies E_j are:

$$\psi_1 = \alpha_A \alpha_B \qquad\qquad E_1/h = +v + \tfrac{1}{4}J$$
$$\psi_2 = \cos\chi\, \alpha_A \beta_B + \sin\chi\, \beta_A \alpha_B \qquad E_2/h = +\tfrac{1}{2}C - \tfrac{1}{4}J$$
$$\psi_3 = -\sin\chi\, \alpha_A \beta_B + \cos\chi\, \beta_A \alpha_B \qquad E_3/h = -\tfrac{1}{2}C - \tfrac{1}{4}J$$
$$\psi_4 = \beta_A \beta_B \qquad\qquad E_4/h = -v + \tfrac{1}{4}J \qquad\qquad (3.4)$$

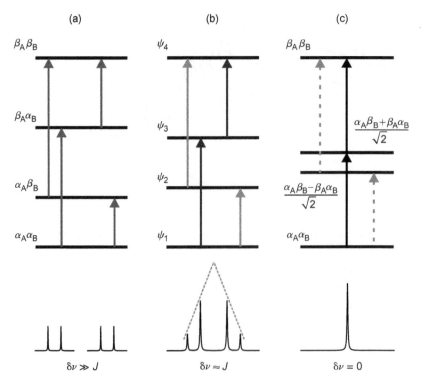

We assume that $\delta v > 0$ and $J > 0$ to avoid the use of moduli, e.g. $|\delta v| \gg |J|$.

Fig. 3.15 Energy levels and spectra of a pair of spin-$\frac{1}{2}$ nuclei, A and B. (a) Weak coupling ($\delta v \gg J$), (b) strong coupling ($\delta v \approx J$), and (c) equivalent spins ($\delta v = 0$). Dashed arrows: forbidden transitions. Solid arrows: the darker the arrow, the higher the transition probability and the stronger the corresponding NMR line.

where

$$v = \tfrac{1}{2}(v_{0A} + v_{0B}), \quad C = \sqrt{J^2 + (\delta v)^2}, \quad \tan 2\chi = \frac{J}{\delta v}. \tag{3.5}$$

As noted above, ψ_2 and ψ_3 are linear combinations of $\alpha_A\beta_B$ and $\beta_A\alpha_B$ and have energies that depend on the strength of the coupling ($J/\delta v$), specified by the angle χ. ψ_1 and ψ_4, being well separated in energy from each other and from ψ_2 and ψ_3, are independent of the coupling strength.

Table 3.3 summarizes the frequencies of the four lines, $(E_j - E_k)/h$, and their relative intensities. In the weak coupling limit ($\delta v \gg J$) it can be seen from eqn 3.5 that $C \approx \delta v$ and $\chi \approx 0$ so that the four lines all have relative intensity 1 and occur at the expected frequencies: $v_{0A} \pm \tfrac{1}{2}J$ and $v_{0B} \pm \tfrac{1}{2}J$. In the other extreme (equivalent spins), $C = J$ and $\chi = 45°$ so that the line positions are $v - J$, v, v, $v + J$ with relative intensities 0, 2, 2, 0, respectively.

The spectra in Fig. 3.15 are generally given the names AX (weak coupling), AB (strong coupling), and A_2 (equivalent spins). The intensity distortions arising from strong coupling are sometimes referred to as the 'roof effect' (indicated by the sloping dashed lines above the spectrum in Fig. 3.15(b)). In the presence of strong coupling, the doublets still have splitting equal to J but they are no longer centred at the chemical shift positions.

Table 3.3 Frequencies and relative intensities of the NMR lines of a strongly coupled pair of $I = \frac{1}{2}$ spins

Transition	Frequency	Relative intensity[a]
$3 \leftrightarrow 4$	$\nu - \frac{1}{2}C - \frac{1}{2}J$	$1 - \dfrac{J}{C}$
$1 \leftrightarrow 2$	$\nu - \frac{1}{2}C + \frac{1}{2}J$	$1 + \dfrac{J}{C}$
$2 \leftrightarrow 4$	$\nu + \frac{1}{2}C - \frac{1}{2}J$	$1 + \dfrac{J}{C}$
$1 \leftrightarrow 3$	$\nu + \frac{1}{2}C + \frac{1}{2}J$	$1 - \dfrac{J}{C}$

[a] When $J > 0$, transitions $3 \leftrightarrow 4$ and $1 \leftrightarrow 3$ are the outer (weaker) lines of each doublet and $1 \leftrightarrow 2$ and $2 \leftrightarrow 4$ are the inner (stronger) lines. ν and C are defined in eqn 3.5.

Closer inspection of eqn 3.4 gives a little more insight into the absence of splittings in the spectra of magnetically equivalent spins. The four states in eqn 3.4 can be classified according to their symmetry with respect to interchange of the A and B labels. When $\delta\nu = 0$, $\psi_3 = 2^{-1/2}(\beta_A\alpha_B - \alpha_A\beta_B)$ and is *antisymmetric* (a *singlet state*). In the same limit, $\psi_1 = \alpha_A\alpha_B$, $\psi_2 = 2^{-1/2}(\alpha_A\beta_B + \beta_A\alpha_B)$ and $\psi_4 = \beta_A\beta_B$, are all *symmetric* (*triplet states*); unlike ψ_3 they do not change sign when the spin labels are exchanged. The three triplet energy levels are equally spaced (Fig. 3.15(c)) and can be thought of as arising from a 'compound' nucleus with $I = 1$. Similarly, the singlet energy level can be regarded as coming from a non-magnetic nucleus ($I = 0$). Looked at in this way, the spectrum of two equivalent spins is simply that of an isolated spin-1 nucleus, i.e. a single line at the chemical shift. Put another way, the two transitions involving the triplet energy levels, $\psi_1 \leftrightarrow \psi_2 \leftrightarrow \psi_4$, are allowed and degenerate, while the singlet–triplet transitions, $\psi_1 \leftrightarrow \psi_3 \leftrightarrow \psi_4$, which would have frequencies $\pm J$ either side of the chemical shift position, are completely forbidden and have zero intensity (Fig. 3.15(c)).

As one might anticipate, the effects of strong coupling can be much more complicated when more than two spins are involved. The multiplet patterns discussed in Section 3.3 can be so severely distorted that they become difficult to recognize; the changes in transition probabilities cause otherwise forbidden transitions to be observed, and chemical shifts and coupling constants can no longer be extracted without a detailed analysis. Such problems are alleviated by the use of high-field spectrometers. Because coupling constants are independent of B_0 and $\delta\nu$ is proportional to B_0 (eqn 2.3), a strongly coupled spin system often becomes weakly coupled at higher field. For example, a pair of protons with $J = 6$ Hz and chemical shift difference of 0.2 ppm would show a pronounced roof effect on a 60 MHz spectrometer ($J/\delta\nu = 0.5$) but not at 600 MHz ($J/\delta\nu = 0.05$).

For discussions of strong coupling effects in larger spin systems, see Bovey (1988) and Günther (2013).

3.6 Mechanism of spin–spin coupling

So far, nothing has been said about the *origin* of spin–spin coupling, apart from some vague statements about nuclei being the source of local magnetic fields that affect the energies of other nuclei. The most obvious interaction between two nearby spins is their mutual *dipolar* coupling (Appendix A). In roughly the same way that two bar magnets interact, so pairs of neighbouring nuclei sense one another's orientation through their dipolar magnetic fields. However, as outlined in Section 3.8, this anisotropic interaction averages to zero for molecules tumbling rapidly and isotropically in solution, and so cannot be responsible for the multiplets discussed in Section 3.3.

The principal source of *J*-coupling in molecules is an *indirect* interaction mediated by the valence electrons.

Contact interaction

We start by considering an electron spin interacting with a nuclear spin. The electron has spin-$\frac{1}{2}$ and a magnetogyric ratio some 660 times that of a proton. Unpaired electrons therefore have strong magnetic dipolar interactions with nearby nuclei but, being purely anisotropic, they should average to zero for molecules tumbling in solution. This indeed happens, *except* at electron–nuclear separations comparable to the nuclear radius ($\sim 10^{-14}$ m) where the particles can no longer be thought of as point dipoles. This breakdown of the point–dipole approximation can be visualized by thinking of the nucleus as a circular current loop of radius $\sim 10^{-14}$ m. Far from the centre of the loop, the field it generates indeed has a $3\cos^2\theta - 1$ dependence (see Section 3.8), but *inside* the loop, the magnetic flux lines are nearly parallel, with little angular variation (see, e.g., Fig. 2.24).

In fact, at very small separations, the dipolar interaction of an electron and a nucleus is replaced by an *isotropic* coupling known as the Fermi *contact interaction*. Its strength is proportional to the scalar product of the two magnetic moments

$$\text{contact interaction} \propto -\gamma_e\gamma_n \, \mathbf{I}\cdot\mathbf{S}, \tag{3.6}$$

where $\mathbf{I}$ and $\mathbf{S}$ are, respectively, the nuclear and electron spin angular momentum vectors. Since the electron has a *negative* magnetogyric ratio ($\gamma_e < 0$), a nucleus with $\gamma_n > 0$ is *stabilized* if the electron and nuclear spins are *antiparallel* ($\mathbf{I}\cdot\mathbf{S} < 0$), and *destabilized* if they are *parallel* ($\mathbf{I}\cdot\mathbf{S} > 0$) (Fig. 3.16). The magnitude of the interaction is also proportional to the probability of finding the electron at the nucleus ($R = 0$) and therefore vanishes unless the electronic wave function has some s-electron character (p, d, f, etc. wave functions have no amplitude at $R = 0$). In short, this isotropic interaction allows an electron spin to sense the state of a nearby nuclear spin, in a way that survives the orientational averaging effect of rapid molecular tumbling.

In paramagnetic atoms and molecules (i.e. those with one or more unpaired electrons), the contact interaction produces *hyperfine* splittings of lines in electronic spectra and electron spin resonance spectra. More importantly in the present context, it provides a pathway for spin–spin coupling between *pairs of nuclei*.

The averaging of dipolar interactions by molecular tumbling is discussed in Sections 3.8 and 5.6.

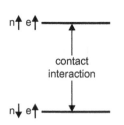

Fig. 3.16 Energy levels of an electron, e, and a spin-$\frac{1}{2}$ nucleus, n ($\gamma_n > 0$), with a Fermi contact interaction. The antiparallel configuration of spins is stabilized relative to the parallel arrangement.

Indirect coupling between nuclei

At first sight it seems unlikely that the contact interaction could form the basis of a general mechanism of nuclear J-coupling. Most molecules have closed electronic shells with no unpaired electrons and therefore, one might think, no contact interactions.

To get an idea of how spin–spin coupling comes about, consider the simplest diamagnetic molecule, H_2. Ignoring normalization constants, the ground state electronic wave function may be written

$$\Psi_0 = \phi_0(\alpha_a\beta_b - \beta_a\alpha_b). \tag{3.7}$$

Ψ_0 has two parts: the spatial wave function ϕ_0 (the molecular orbital) and the electron-spin function. The two electron spins, a and b, are paired (i.e. a singlet state) in a bonding orbital formed from the two atomic 1s orbitals (Fig. 3.17(a)). As before, α and β are shorthand for $m = +\frac{1}{2}$ and $m = -\frac{1}{2}$, respectively. From the form of Ψ_0 and the Born interpretation of the wave function, it is clear that the spatial distributions of the α and β states of both electrons are identical and given by $|\phi_0|^2$.

The contact interaction mixes the singlet ground state with electronically excited triplet states of the molecule. Crudely speaking, this happens because the nucleus–electron coupling can flip the spin of one of the electrons, converting singlet (antiparallel spins) to triplet (parallel spins), while simultaneously flipping the nuclear spin in the opposite sense so as to conserve angular momentum. The singlet → triplet mixing must be accompanied by electronic excitation because the Pauli principle forbids two electrons with parallel spins to be in the same orbital. In the case of H_2, the lowest excited triplet state is accessed by promoting one of the two electrons from the bonding orbital into an antibonding orbital (shown in Fig. 3.17(b)). The wave function of this excited state is

$$\Psi_1 = \phi_1(\alpha_a\beta_b + \beta_a\alpha_b), \tag{3.8}$$

which has a symmetric spin part (we ignore the other two triplet spin functions, $\alpha_a\alpha_b$ and $\beta_a\beta_b$, to keep things simple) and an antisymmetric spatial part ϕ_1, which differs from ϕ_0 because of the antibonding contribution. Mixing of the singlet and triplet states by the contact interaction causes the molecular wave function to be a linear combination of Ψ_0 and Ψ_1 (again ignoring normalization constants):

$$\Psi = \Psi_0 + \lambda\Psi_1 = (\phi_0 + \lambda\phi_1)\alpha_a\beta_b - (\phi_0 - \lambda\phi_1)\beta_a\alpha_b, \tag{3.9}$$

where λ is a small constant determined by the strength of the contact interaction and the energy of the excited state Ψ_1 above the ground state Ψ_0. Since ϕ_0 and ϕ_1 have different shapes (Fig. 3.17), the probability of finding electron a with spin α_a at a given position in the molecule $(\sim|\phi_0 + \lambda\phi_1|^2)$ differs from the corresponding probability for β_a $(\sim|\phi_0 - \lambda\phi_1|^2)$. The electronic wave function has become *spin-polarized* (Fig. 3.18).

It is now straightforward to see how this leads to an interaction between the two protons (Fig. 3.19). If proton A has spin β, the spin polarization leads to a slight excess of α electron spins and a slight depletion of β electron spins in its vicinity

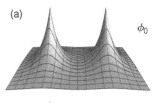

(a)
ϕ_0

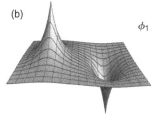

(b)
ϕ_1

Fig. 3.17 Representations of the bonding (ϕ_0) and antibonding (ϕ_1) molecular orbitals of H_2 (see eqns 3.7 and 3.8).

This description of the origin of spin-spin coupling is a simplified version of one to be found in Carrington and McLachlan (1967).

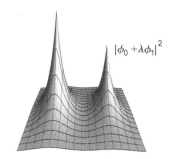

$|\phi_0 + \lambda\phi_1|^2$

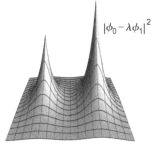

$|\phi_0 - \lambda\phi_1|^2$

Fig. 3.18 Spin-polarized molecular orbitals of H_2 resulting from Fermi contact interactions. The sketches show the probability of finding an electron in its α spin state ($|\phi_0 + \lambda\phi_1|^2$) and its β spin state ($|\phi_0 - \lambda\phi_1|^2$). The degree of spin polarization has been greatly exaggerated.

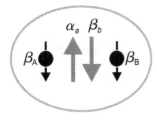

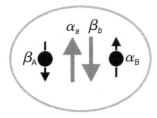

Fig. 3.19 ^{1}H-^{1}H J-coupling in H_2. Nuclear spins, A and B, are shown as black circles and arrows. Electron spins, a and b, are shown as grey arrows. The lower part of the figure shows the low energy configuration in which the nuclear spins are antiparallel. The upper part shows the high energy configuration with parallel nuclear spins.

(remember that the contact interaction stabilizes antiparallel electron and proton spins). There is a corresponding build-up of β electron spins and reduction of α electron spins near the other proton, B. If B has spin α, it will be stabilized by the local excess of β electron spins through its contact interaction (Fig. 3.18). Conversely, if proton B has spin β, it will be destabilized. In this way one nucleus senses the spin of the other via the valence electrons. If the spin of proton A is inverted, the situation is reversed, and there is a small accumulation of α electron spins around proton B, which is consequently stabilized when it has spin β.

Similar arguments can be used to rationalize the existence of spin–spin interactions in larger molecules. Generally speaking, the strength of the coupling falls off rapidly as the number of intervening bonds increases. In reality, the mechanism of J-coupling is rather more complex than suggested by the simple-minded model presented above. Just as with chemical shifts (Section 2.4), it is now possible to calculate J-couplings, fairly reliably in many cases, using the methods of *ab initio* quantum chemistry (e.g. Bonhomme et al. (2012)). However, in the following paragraphs we discuss a few examples of cases in which J-couplings can be related simply and qualitatively to molecular and electronic structure.

3.7 Properties of J-coupling

The highly simplified arguments of the previous section give an impression of the mechanism of spin–spin coupling and indicate its general properties. The strength of the interaction is crucially dependent on the s-character of the wave functions of the ground state and electronically excited states at the positions of the nuclei. The coupling is not affected by the strength of the external magnetic field, in contrast to the differences in resonance frequencies that arise from chemical shifts. J-couplings are therefore independent of the spectrometer frequency and, being isotropic, are not affected by molecular tumbling.

One-bond and two-bond couplings

The interpretation of the magnitudes of J-coupling constants is, in most cases, even more of a problem than it is for chemical shifts, and not one that will be tackled here. Instead, a few representative coupling constants are summarized (Figs 3.20, 3.21, 3.23, and 3.26) together with the briefest of comments.

One-bond carbon–proton couplings ($^1J_{CH}$) generally fall in the range 100–250 Hz, and are sensitive to the s-electron character of the carbon atomic orbital involved in the C–H bond, reflecting the crucial role played by the contact interaction. The hydrocarbons ethane, ethylene, and acetylene, which have, respectively, sp^3, sp^2, and sp hybridization, obey the empirical relation:

$$^1J_{CH} / Hz \approx 5 \times \%(s), \tag{3.10}$$

where %(s), the percentage s-character of the C–H bond, equals 25, 33, and 50, respectively (Fig. 3.20). Similar effects of hybridization are found for strained rings (Fig. 3.20): the smaller the ring size the larger the p-character of the C–C bonds in the ring,

H$_3$C—CH$_3$	125	CH$_4$	125	CH$_3$Cl	147
H$_2$C=CH$_2$	157	CH$_3$OH	141	CH$_2$Cl$_2$	177
HC≡CH	250	CH$_3$CN	136	CHCl$_3$	208

| 123 | 128 | 136 | 161 | 205 |

Fig. 3.20 One-bond ^{13}C-^{1}H coupling constants (in Hz).

X	H$_2$C=CHX	CH$_3$—X
H	+2.3	−12.4
Ph	+1.3	−14.5
Cl	−1.3	−10.8
CN	+0.9	−16.9

Fig. 3.21 Two-bond ^{1}H-^{1}H coupling constants (in Hz).

and consequently the larger the s-character of the carbon orbitals used to form the C–H bonds. Figure 3.20 also gives a few examples illustrating the effect of substituents.

Two-bond (geminal) proton–proton couplings vary over a wide range (approximately, −20 to +40 Hz) with large substituent effects; sp^2 hybridized CH$_2$ groups generally have smaller $^2J_{HH}$ than do methyl groups (Fig. 3.21).

Three-bond couplings

Probably the most useful J-couplings are those involving nuclei separated by *three* bonds, for example $^3J_{HH}$ in an H–C–C–H fragment. Experimentally and theoretically, these coupling constants are found to vary with the dihedral angle between the two H–C–C planes (θ, see Fig. 3.22) according to the 'Karplus relation':

$$^3J \approx A + B\cos\theta + C\cos^2\theta. \tag{3.11}$$

Although it is possible to calculate approximate values for A, B, and C (including substituent and other effects), it is more satisfactory to treat them as coefficients to be determined empirically using conformationally rigid model compounds of known structure. Typical values are $A = 2$ Hz, $B = -1$ Hz, $C = 10$ Hz, which give a θ-variation of the type shown in Fig. 3.22 (a 'Karplus curve'). The values of the three parameters depend on the substituents on the carbon atoms.

The utility of three-bond couplings lies principally in conformational analysis: $^3J_{HH}$ values for the ring protons in cyclohexanes depend on whether axial or equatorial protons are involved; and the *trans* ^{1}H-^{1}H couplings across a C=C bond are up to a factor of two larger than the *cis* couplings (Fig. 3.23).

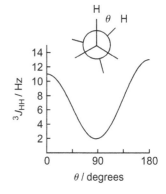

Fig. 3.22 Typical dependence of a three-bond H–C–C–H coupling constant on the dihedral angle θ.

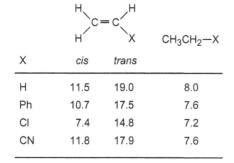

X	cis	trans	CH$_3$CH$_2$—X
H	11.5	19.0	8.0
Ph	10.7	17.5	7.6
Cl	7.4	14.8	7.2
CN	11.8	17.9	7.6

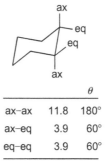

		θ
ax–ax	11.8	180°
ax–eq	3.9	60°
eq–eq	3.9	60°

Fig. 3.23 Three-bond ^{1}H-^{1}H coupling constants (in Hz).

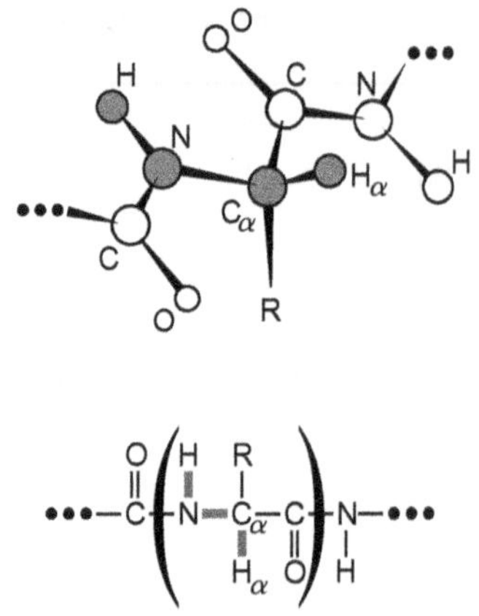

Fig. 3.25 The three staggered conformations of an amino acid shown in Newman projection with C_α in front and C_β behind.

Fig. 3.24 Part of the backbone of a polypeptide chain, showing the H–N–C_α–H dihedral angle. R is the side-chain of the amino acid residue shown in brackets in the lower part of the figure.

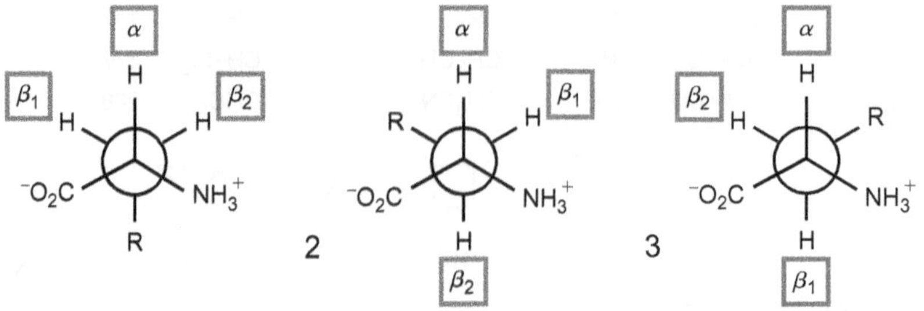

The Karplus relation finds valuable applications in studies of protein structures. For example, the couplings between the amide (NH) and C_α protons in a polypeptide chain provide information on the conformation of the protein backbone (Fig. 3.24). In particular, the two major elements of secondary structure in proteins—α-helices and β-sheets—have characteristic H–N–C_α–H dihedral angles: ~120 and ~180°, respectively. Thus, $^3J_{HH}$ values smaller than 6 Hz often indicate an α-helix, while couplings larger than about 7 Hz generally arise from β-sheet regions of the protein.

The interpretation of three-bond couplings in conformationally mobile molecules is somewhat different. Consider, for example, the coupling between the α-proton and the two β-protons in an amino acid (Fig. 3.25). The three staggered conformations, or rotamers, interconvert rapidly so that the two observed $^3J_{\alpha\beta}$ values are averages, weighted according to the populations of the three energy minima:

$$J_{\alpha\beta_1} = P_1 J_g + P_2 J_g + P_3 J_t$$
$$J_{\alpha\beta_2} = P_1 J_g + P_2 J_t + P_3 J_g$$

(3.12)

where $P_1 + P_2 + P_3 = 1$. J_t and J_g are *trans* ($\theta = 180°$) and *gauche* ($\theta = \pm 60°$) three-bond coupling constants. Substituent effects on J_t and J_g are ignored here for simplicity. The relative populations of the three rotamers can therefore be determined provided J_t and J_g are available from measurements on rigid model compounds or calculations.

Long-range couplings

Proton–proton coupling constants are generally very small (<1 Hz) when the nuclei are separated by more than three bonds. A few of the exceptions are shown in Fig. 3.26. Note that large $^4J_{HH}$ and $^5J_{HH}$ often occur when the coupling is transmitted along a zigzag arrangement of bonds and/or through π-bonds.

Fig. 3.26 Long range ^{1}H–^{1}H coupling constants (in Hz).

3.8 Dipolar coupling

Finally, we turn to the direct dipolar interactions between nuclei which, though not normally responsible for splittings in the spectra of molecules in the liquid state, are important in solid-state NMR and for spin relaxation (Chapter 5). The basic features of dipolar interactions are presented in Appendix A.

Dipolar interaction in solids

Equation A.2 in Appendix A gives an expression for the energy of interaction of two classical magnetic moments, $\boldsymbol{\mu}_A$ and $\boldsymbol{\mu}_X$, both pointing along the positive z-axis:

$$E = -\left(\frac{\mu_0}{4\pi}\right)\left(\frac{\mu_A\mu_X}{r^3}\right)(3\cos^2\theta - 1), \tag{3.13}$$

where r is the internuclear distance, θ is the angle between the internuclear vector and the z-axis, and $\mu_0 = 4\pi \times 10^{-7}$ H m^{-1} is the vacuum permeability.

We consider first the *heteronuclear* case in which A and X are spin-$\frac{1}{2}$ nuclei with different magnetogyric ratios ($\gamma_A \neq \gamma_X$). In the strong magnetic field of an NMR spectrometer, both spins are quantized along the field direction (the z-axis). To make eqn 3.13 applicable to nuclear spins (i.e. quantum rather than classical magnetic moments), we can just replace μ_A and μ_X by their z-components, $\gamma_A m_A\hbar$ and $\gamma_X m_X\hbar$, respectively (using $\mu_z = \gamma I_z$ and $I_z = m\hbar$ as in Chapter 1). This gives

$$E = -hR_{AX}(3\cos^2\theta - 1)m_Am_X, \tag{3.14}$$

where

$$R_{AX} = \left(\frac{\hbar}{2\pi}\right)\left(\frac{\mu_0}{4\pi}\right)\left(\frac{\gamma_A\gamma_X}{r^3}\right) \tag{3.15}$$

is the dipolar coupling constant (in Hz). Comparing eqn 3.14 with the corresponding expression for a pair of weakly J-coupled spins (eqn 3.1), one can see that the NMR signals of A and X will both be doublets with a splitting

$$R_{AX}(3\cos^2\theta - 1). \tag{3.16}$$

For example, if the two nuclei are ^{1}H and ^{13}C, $R_{CH} = 8951$ Hz when $r = 1.5$ Å; 472 Hz at 4 Å; and 30 Hz at 10 Å. Compared to J-couplings, dipolar interactions are strong and long range.

Figure 3.27 shows NMR spectra calculated for a range of values of θ between 0° and 90°. These are the sort of spectra that would be observed for isolated AX pairs in a single crystal as the crystal is rotated in the magnetic field of the spectrometer. As θ changes from 0° ($3\cos^2\theta - 1 = 2$) to 90° ($3\cos^2\theta - 1 = -1$) the doublet splitting decreases, goes through zero at 54.7° (the so-called *magic angle*) and then increases again as θ rises to 90°. Identical behaviour is found in both the A and X spectra. Internuclear separations may easily be determined from single crystal spectra of such simple spin systems.

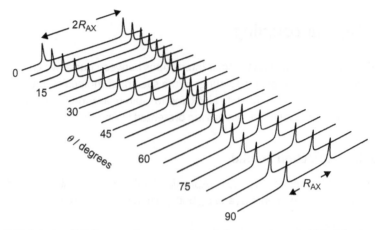

Fig. 3.27 Calculated NMR spectra for one member of a heteronuclear pair (AX) of dipolar-coupled spin-$\frac{1}{2}$ nuclei. θ is the angle between the internuclear vector and the magnetic field direction. The dipolar coupling constant R_{AX} is given by eqn 3.15.

The situation is a bit more complicated for a powdered sample. Although each AX pair has a unique value of θ, different molecules have different θ. Assuming a random distribution of orientations, the observed 'powder spectrum' is the sum of the single crystal spectra for θ between 0 and 90°, each weighted by $\sin\theta$ to take into account the probability of finding an AX pair with orientation θ. Adding these spectra together produces the unusual lineshape shown in Fig. 3.28: the 'horns' correspond to $\theta \approx 90°$, while the wings come from the $\theta \approx 0°$ orientations.

As may be anticipated, both single crystal and powder spectra are somewhat more complicated for larger spin systems, where each nucleus may have significant dipolar interactions with many neighbouring spins, each with its own r and θ.

Homonuclear dipolar couplings, between spins with the same γ, produce single crystal and powder spectra that are essentially identical to those arising from heteronuclear interactions. The only difference is that the doublet splitting contains an extra factor of $\frac{3}{2}$:

$$\tfrac{3}{2}R_{AX}\left(3\cos^2\theta - 1\right) \quad \text{and} \quad R_{AX} = \left(\frac{\hbar}{2\pi}\right)\left(\frac{\mu_0}{4\pi}\right)\left(\frac{\gamma^2}{r^3}\right). \tag{3.17}$$

Briefly, the $\frac{3}{2}$ arises because the homonuclear dipolar interaction mixes the spin states $\alpha_A\beta_X$ and $\beta_A\alpha_X$. This does not occur in the heteronuclear case because the energy gap between $\alpha_A\beta_X$ and $\beta_A\alpha_X$ ($\approx \hbar|\gamma_A - \gamma_X|B_0$) is much greater than the strength of the coupling ($\approx hR_{AX}$). The effect is very similar to the mixing that occurs for strongly J-coupled spins (Section 3.5); however the form of the dipolar coupling changes the details of the mixing such that dipolar splittings *are* observed for equivalent nuclei. For example, the ^{1}H NMR spectrum of an isolated water molecule in a crystal is a doublet with splitting given by eqn 3.17 ($R_{HH} = 30.5$ kHz for $r = 1.58$ Å).

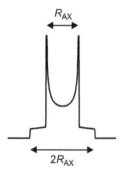

Fig. 3.28 Calculated powder spectrum for one member of a heteronuclear pair of dipolar-coupled spin-$\frac{1}{2}$ nuclei. This lineshape is often referred to as a 'Pake pattern'.

Dipolar interaction in liquids

Evidently dipolar couplings have a profound effect on the NMR spectra of solids, but what about liquids with which this book is principally concerned?

Molecules in liquids rotate rapidly with frequent changes in the axis and speed of rotation as a result of collisions with other molecules. Consequently, for any pair of nuclei, the angle θ and therefore their dipolar interaction are rapidly modulated. As discussed in more detail in Section 5.6, this leads to an *average* splitting, provided the average rotational frequency greatly exceeds the strength of the dipolar coupling. This condition is certainly met for all but very large molecules and/or very viscous solutions. For example, a water molecule at room temperature has a rotation frequency of ~10^{12} Hz, while the largest dipolar couplings are no more than 10^5 Hz. To obtain the dipolar splitting of a molecule in a liquid, the angular parts of eqns 3.16 and 3.17 must therefore be averaged over all molecular orientations, θ:

$$\text{splitting} = \lambda R_{AX} \int_0^{\pi/2} \left(3\cos^2\theta - 1\right)\sin\theta \, d\theta, \tag{3.18}$$

where $\sin\theta$ is the appropriate weighting factor for a molecule that has *no preferred orientation* and λ equals 1 for heteronuclear spins (eqn 3.16) and $\frac{3}{2}$ for homonuclear spins (eqn 3.17). This integral is identically zero: the positive parts of the integrand ($0 \le \theta \le 54.7°$) exactly cancel the negative parts ($54.7° \le \theta \le 90°$). Thus, dipolar interactions do not normally produce splittings in the NMR spectra of liquids. (But they do play a crucial role in *spin relaxation*, as we shall see in Chapter 5.)

However, the $\sin\theta$ weighting factor in eqn 3.18 is not always appropriate. When molecules are partially aligned with respect to the magnetic field B_0, the average dipolar splitting does not vanish because the positive and negative parts of the integrand no longer exactly cancel. This can come about when the interaction of the molecules with B_0 is *anisotropic* such that certain molecular orientations have lower energy and are thus more prevalent than others. Alternatively, molecules can be partially aligned if they are dissolved in a medium that is itself aligned with B_0. Examples of such media include solutions of rod-shaped viruses or disc-shaped assemblies of phospholipids (known as bicelles) and stretched or compressed polyacrylamide gels (Cavanagh et al. (2007)). Under such conditions, the measured spin–spin coupling of two nuclear spins will be

Molecules tend to orient in a magnetic field when their magnetic susceptibilities are significantly anisotropic, i.e. when the field induces a magnetic moment in the molecule that depends on its orientation. Molecules containing paramagnetic metal atoms often have large anisotropic magnetic susceptibilities.

$$\text{splitting} = J_{AX} - \lambda R_{AX} \left\langle 3\cos^2\theta - 1\right\rangle, \tag{3.19}$$

where J_{AX} is the J-coupling and the angled brackets indicate the orientational average. These *residual dipolar couplings* provide valuable information on the orientation of internuclear vectors within molecules. The idea is to tune the degree of molecular alignment such that only the spin–spin couplings (eqn 3.19) of directly bonded atoms are affected. Knowing the bond lengths, it is then possible to determine the relative orientations of, for example, all the backbone amide ^1H–^{15}N bonds within a protein. When allied with nuclear Overhauser enhancements (Section 5.8) and torsion angles from three-bond J-couplings (Section 3.7), residual dipolar couplings provide a powerful method for protein structure determination (see e.g. Kwan et al. (2011)).

The properties of J- and dipolar couplings are summarized in Table 3.4.

Table 3.4 Properties of J-couplings and dipolar couplings

J-couplings	Dipolar couplings
Through bonds	Through space
Strength^a < 100 Hz	Strength^a < 100 kHz
Isotropic	$3\cos^2\theta - 1$
Small for > 3 bonds^a	$1/r^3$
Cause splitting in spectra^b	Splitting for solids and partially aligned molecules in liquids.
Do not cause relaxation^c	Cause relaxation when motion present

^{a} The strengths of these interactions depend on the magnetogyric ratios of nuclei involved and on molecular structure. The numbers are very approximate.
^{b} Unless averaged by chemical exchange.
^{c} Unless modulated by internal motion.

3.9 Summary

- Spin–spin interactions (J-couplings) give rise to multiplets in liquid-state NMR spectra.

- J-coupling usually occurs via chemical bonds and is generally small when the nuclei are more than three bonds apart.

- J-couplings give information on bonding networks in molecules.

- J-couplings within a group of magnetically equivalent spins do not produce multiplet splittings.

- Strongly coupled spins have more complex spectra than weakly coupled spins.

- Dipolar couplings occur through space and normally do not lead to multiplets in liquid-state NMR spectra.

3.10 Exercises

Answers to the exercises are provided at the back of the book. Full worked solutions are available on the Online Resource Centre at www. oxfordtextbooks.co.uk/orc/hore_nmr2e/

1. Suggest structures for the following compounds based on the multiplets observed in their NMR spectra (ignore spin–spin couplings involving Cl and I): (a) ^{19}F spectrum of ClF$_3$: doublet and triplet. (b) ^{19}F spectrum of IF$_5$: doublet and quintet. (c) ^1H spectrum of C$_3$H$_7$Cl: doublet and septet. (d) ^1H spectrum of C$_2$H$_3$OCl: three doublets of doublets. (e) The ^{51}V spectrum of VOF$_4^-$ is a quintet. The ^{19}F spectrum comprises eight equally spaced lines with the same intensity. Propose a structure for VOF$_4^-$ and determine the spin quantum number of ^{51}V.

2. Predict the total number of lines in the ^1H spectra of the following compounds: (a) CH$_3$Cl. (b) (CH$_3$)$_3$CH. (c) 1,4-dichloro-2,3-dibromobenzene. (d) 1,2-dichloro-3,4-dibromobenzene. (e) 1,1-dichlorocyclopropane. Assume that all J-couplings involving Br and Cl are negligibly small.

3. The ^{1}H spectrum of CH_2D_2 contains five lines. What are their relative intensities? (b) How many lines are there in the ^{1}H spectrum of CHD_3?

4. The ^{1}H spectra of which isomers of C_4H_9Cl contain the following multiplets? (a) doublet, triplet, quintet, and sextet; (b) triplet, triplet, quintet, and sextet. Assume all three-bond J_{HH} are identical and ignore all other couplings.

5. What would the spectrum in Fig. 3.13 look like if all the lines had widths of ~3 Hz?

6. The ^{1}H spectrum of an AX spin system has lines at the following frequencies: 600.001677, 600.001683, 600.004437, 600.004443 MHz. Taking the Larmor frequency of TMS to be exactly 600 MHz, determine the two chemical shifts and the J-coupling.

7. For each of the following compounds determine whether the protons are magnetically or chemically equivalent. (a) benzene. (b) the 2,5 protons in furan. (c) $F_2C=C=CH_2$. (d) $H-^{13}C\equiv^{13}C-H$.

8. The protons in 2-bromo-5-chlorothiophene have a chemical shift difference of 0.154 ppm and a J-coupling of 3.9 Hz. Determine the ratio of the intensities of the inner and outer lines of the four-line ^{1}H spectrum on (a) a 600 MHz and (b) a 40 MHz spectrometer.

9. The ^{1}H-^{1}H J-coupling in a compound CHX_2-CHY_2 is 3.46 Hz. If $J_g = 2.2$ Hz and $J_t = 9.7$ Hz, determine the mole fractions of the two rotamers.

10. ^{13}C NMR spectra of a single crystal of isotopically enriched glycine, $^{15}NH_3^+-^{13}CH_2-^{13}CO_2^-$, were measured for different orientations of the crystal. The maximum ^{15}N-$^{13}CH_2$ and ^{13}C-^{13}C dipolar splittings were found to be 1.941 kHz and 6.414 kHz, respectively. Determine (a) the C–C and (b) the C–N bond lengths.

4 Chemical exchange

4.1 Introduction

The two previous chapters have dealt with the interactions that give rise to NMR lines. We now turn to processes capable of removing, or at least modifying, some of this structure, namely *dynamic equilibria*.

We start by considering the simplest case of dynamic equilibrium: a molecule interconverting between two conformations of equal energy:

$$A \underset{k}{\overset{k}{\rightleftharpoons}} B \qquad\qquad (4.1)$$

with identical forward and backward first-order rate constants, k. A good example is dimethylnitrosamine (Fig. 4.1). The skeleton of this molecule is planar, due to the partial double-bond character of the N–N bond, except for brief instants when the nitroso group undergoes 180° flips which interconvert the two degenerate forms (Fig. 4.1). At low temperatures, the internal rotation is slow and the ^{1}H spectrum comprises two equally intense resonances from the methyl groups *cis* and *trans* to the oxygen with Larmor frequencies $v_{0,cis}$ and $v_{0,trans}$ (the coupling between the two sets of protons is too small to be resolved). At higher temperatures, the nitroso group flips at an appreciable rate, and every time it does so the chemical shifts of the two methyl groups are interchanged. The Larmor frequency of each group of protons thus hops from $v_{0,cis}$ to $v_{0,trans}$ and back again, with an average time between jumps of $\tau = 1/k$. Processes such as this are known as *chemical exchange*, even though (as here) there may be no making or breaking of bonds.

The theory of chemical exchange is pleasingly straightforward compared to the complex computations required to obtain chemical shifts and *J*-couplings. Nevertheless, we shall avoid the algebra here (see, for example, Carrington and McLachlan (1967) and Günther (2013)): the formulae involved are rather cumbersome, give little physical insight, and are not essential for a qualitative appreciation of the origin of the effects and how they may be used to obtain chemical information. Instead, we summarize the results for a couple of simple cases, discuss their origin in hand-waving terms, and present a variety of examples and applications.

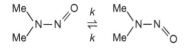

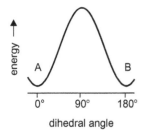

Fig. 4.1 Dimethylnitrosamine as an example of symmetrical two-site exchange. The two conformations of the molecule have identical energy and interconvert by 180° rotations of the N=O group around the N–N bond, which has partial double-bond character. The transition state corresponds to a C–N–N–O dihedral angle of 90°.

4.2 Symmetrical two-site exchange

For the moment we stay with the simple two-site exchange A $\rightleftharpoons$ B, with equal forward and backward rate constants. Figure 4.2 shows a set of spectra calculated for a range of rate constants, k. ν_{0a} and ν_{0b} are the Larmor frequencies when $k=0$. For very slow exchange, one sees two equally intense, narrow peaks at ν_{0a} and ν_{0b}. As k is increased, the two lines first broaden, then shift towards one another, broadening further until they merge into a single, wide, flat-topped line. This happens when k is similar in magnitude to the *difference* in resonance frequencies, $\delta\nu=\nu_{0a}-\nu_{0b}$ of the two exchanging sites. Further increase in the exchange rate produces a sharp resonance at the mean frequency, $\frac{1}{2}(\nu_{0a}+\nu_{0b})$. Thus, if the exchange is fast enough the difference in resonance frequencies of the two sites collapses to zero. Experimentally, such changes in the spectrum may be observed by increasing the temperature.

We assume $\delta\nu > 0$ so as to avoid using $|\delta\nu|$.

Several questions immediately arise. Why are the two lines broadened by slow exchange? Why do they merge into a single sharp line when the exchange is fast rather than continuing to get broader and broader as k increases? And why does the ratio $k/\delta\nu$ determine whether the two environments are averaged or not? The following sections attempt to provide answers.

Slow exchange

Slow exchange is the regime in which the separate resonances are exchange-broadened but still to be found at frequencies ν_{0a} and ν_{0b} (see Fig. 4.2). In this

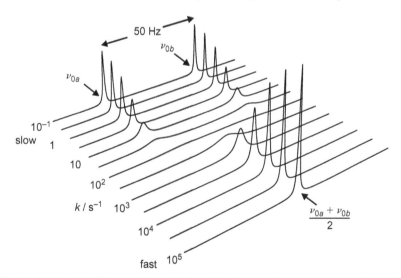

Fig. 4.2 Calculated NMR spectra for a pair of nuclei exchanging between two sites with equal populations (symmetrical two-site exchange). Spectra are shown for a range of values of the exchange rate constant k. The difference in resonance frequencies of the two sites, $\delta\nu$, is 50 Hz. The linewidths in the absence of exchange are 1 Hz.

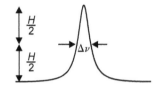

Fig. 4.3 Δv is defined as the full width of the NMR line at half its maximum height. In this chapter, Δv refers to the line broadening arising from exchange processes, i.e. the total linewidth less that in the absence of exchange.

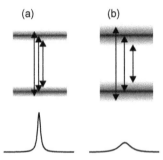

Fig. 4.4 Schematic representation of the blurring of energy levels and broadening of NMR lines brought about by slow chemical exchange. The exchange rate constant is faster in (b) than in (a).

limit, the increase in linewidth (in Hz) as a result of exchange is simply given by

$$\Delta v = \frac{k}{\pi} = \frac{1}{\pi\tau},$$ (4.2)

i.e. the faster the exchange, the wider the line (Δv is defined in Fig. 4.3). The origin of this effect is *lifetime broadening*, sometimes called *uncertainty broadening* because of its loose connection with Heisenberg's uncertainty principle. The energy of a state of finite lifetime cannot be specified precisely—the shorter lived the state, the greater the imprecision in its energy—and, as indicated in Fig. 4.4, transitions between such 'blurred' energy levels result in broadened spectroscopic lines. It is for this reason that electronic spectra have larger natural linewidths than do rotational or vibrational spectra (faster spontaneous emission), and that microwave spectra broaden at high pressure (faster collisional deactivation of excited states).

Other contributions to NMR linewidths are discussed in Chapter 5.

Fast exchange

The other extreme, in which the two lines have merged to form a single, broadened line at the mean Larmor frequency, is termed fast exchange. In this limit, the extra linewidth due to chemical exchange is

$$\Delta v = \frac{\pi(\delta v)^2}{2k} = \frac{\pi(\delta v)^2 \tau}{2},$$ (4.3)

where, as before, $\delta v = v_{0a} - v_{0b}$. In contrast to slow exchange, where the separate resonances are broadened by site-hopping, the single line observed in fast exchange becomes *narrower* as the rate constant increases, reflecting the more efficient averaging of the two environments. That is, fast exchange causes the spins to experience an effective local field that is the mean of the local fields at the two exchanging sites.

To be able to detect *separate* signals from the two sites, they must acquire an appreciable *phase difference*, say 180°, which takes a time $\frac{1}{2}(\delta v)^{-1}$ (Fig. 4.5(a)). Now, if exchange occurs during this period, the accumulation of the phase difference is interrupted. When the spins swap frequencies, δv changes sign, and the phase difference starts to decrease. When the spins jump back to their original frequencies, their phase difference increases once more. So, as shown in Fig. 4.5(b), the phase difference undergoes a random walk with frequent reversals. The result, in the fast exchange regime, is a difference at the end of the $\frac{1}{2}(\delta v)^{-1}$ period that is much smaller than 180° so that, *in effect*, the two signals have very similar frequencies. In other words, fast exchange destroys the frequency difference δv, provided the site-hopping is much faster than the build-up of the phase difference. The resonance observed in the fast exchange limit appears at the mean frequency $\frac{1}{2}(v_{0a} + v_{0b})$, because each spin spends, on average, 50% of its time in each site.

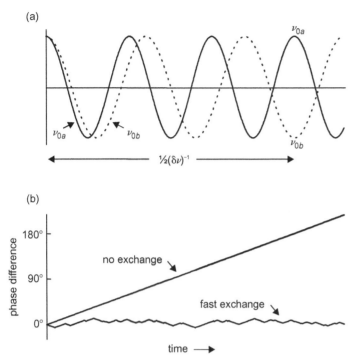

Fig. 4.5 (a) Accumulation of a phase difference between two spins with Larmor frequencies v_{0a} and v_{0b}. (b) Time dependence of the phase difference in the absence of exchange, and under conditions of fast exchange.

Intermediate exchange

As its name implies, intermediate exchange fills the gap between slow and fast exchange. The condition for the two resonances *just* to merge into a single broad line—the point at which the valley between the two peaks just disappears—is

$$k_{merge} = \frac{\pi \delta v}{\sqrt{2}} \approx 2.2 \delta v. \tag{4.4}$$

When $k > k_{merge}$, a single line is expected at the mean resonance frequency; when $k < k_{merge}$, two separate resonances should be seen (Fig. 4.2). NMR signals can be very broad in the intermediate regime and may be difficult to detect if the signal-to-noise ratio (see Section 6.3) is low.

NMR timescale

Whether the exchange is slow, intermediate, or fast is determined by the size of the exchange rate constant k relative to the frequency difference δv, as shown by the discussion of phase differences above. This can also be seen from eqn 4.4. That is, the *timescale* of these events is governed by $(\delta v)^{-1}$. A process described as 'slow, or fast, on the NMR timescale' is slow, or fast, *compared to the difference in*

NMR frequencies of the exchanging sites. Since δv values are rarely larger than ~10 kHz, and often much smaller, only relatively slow processes (seconds to microseconds) can be studied by NMR.

For example, on a 100 MHz spectrometer, the resonances of two protons with a chemical shift difference of 1 ppm would merge when $k=220$ s^{-1}, i.e. when the average lifetime of the two sites is 4.5 ms. On a 500 MHz spectrometer, however, the exchange rate would need to be 1100 s^{-1} (910 μs lifetime) to cause the same two lines to coalesce. Somewhat faster processes can often be studied by ^{13}C NMR because of the larger spread in resonance frequencies. Although the magnetogyric ratio of ^{13}C is a quarter that of ^{1}H, ^{13}C chemical shifts cover a ppm range about 20 times greater, giving a range of Larmor frequencies larger by a factor of five.

4.3 Unsymmetrical two-site exchange

Two-site exchange with equal populations is clearly a special case of the more general, and common, situation in which the two sites can have arbitrary concentrations. For example, if one of the methyl groups in dimethylnitrosamine is replaced by a benzyl group, the two conformers no longer have the same energy; the concentrations differ and so do the forward and backward rate constants (Fig. 4.6). Nevertheless, 180° flips of the nitroso group still modulate the Larmor frequencies (v_{0a} and v_{0b}) of the methyl protons in the two conformers.

If the fractional populations of the two sites are p_A and p_B ($p_A+p_B=1$) and the two first-order rate constants are k_A and k_B:

$$A \underset{k_B}{\overset{k_A}{\rightleftharpoons}} B, \tag{4.5}$$

then $p_A k_A = p_B k_B$ at equilibrium. The average lifetimes of the two sites are $\tau_A = 1/k_A$ and $\tau_B = 1/k_B$.

In the slow exchange limit, the lifetime broadenings are

$$\Delta v_A = \frac{k_A}{\pi} = \frac{1}{\pi\tau_A} \quad \text{and} \quad \Delta v_B = \frac{k_B}{\pi} = \frac{1}{\pi\tau_B}. \tag{4.6}$$

Notice that the two resonances are unequally affected because of their different lifetimes (the less concentrated species has the shorter lifetime and the larger broadening). For fast exchange, a single resonance is observed at the *weighted average frequency*

$$v_{av} = p_A v_{0a} + p_B v_{0b} \tag{4.7}$$

with a linewidth

$$\Delta v = \frac{4\pi p_A p_B (\delta v)^2}{k_A + k_B}. \tag{4.8}$$

All these expressions reassuringly reduce to the formulae for symmetrical exchange when $p_A = p_B = \frac{1}{2}$ and $k_A = k_B = k$. Equation 4.7 should not be surprising: each nucleus spends a proportion p_A of its time in site A, and p_B in site B, giving

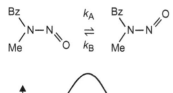

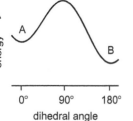

Fig. 4.6 Benzylmethylnitrosamine as an example of unsymmetrical two-site exchange. Bz ≡ $C_6H_5CH_2$.

an average resonance frequency weighted in favour of the more stable site. As a result, ν_{av} depends on the relative amounts of A and B which can be changed, for example, by altering the temperature.

Figure 4.7 shows spectra calculated for a range of exchange rate constants for the case where B has twice the population of A (i.e. $k_A/k_B=2$). As the rate constants are increased, one sees the same pattern of broadening, coalescence, and subsequent narrowing as found for symmetrical exchange (Fig. 4.2).

The principles established for exchange between two sites may be applied to multiple-site exchange processes. The spectra are, not surprisingly, rather more complicated and generally need to be analysed by computer. Typically, a reaction scheme is proposed, spectra are calculated, and the rate constant(s) varied to obtain a match between simulated and experimental spectra.

Processes other than exchange are capable of averaging resonance frequency differences. The most important instances are the averaging of dipolar (Section 3.8) and quadrupolar (Section 5.7) splittings by molecular tumbling in solution. The condition for averaging is the same as for fast exchange: the rate at which the resonance frequencies are exchanged must be fast compared to the frequency differences.

So far only *conformational* equilibria have been mentioned. As we shall see, similar and in some cases identical effects are found for *chemical* equilibria in which bonds are broken and formed. Other aspects of chemical exchange, and the variety of processes that can be studied, are best revealed by a few examples. Several instances of chemical exchange have already been encountered in previous chapters: e.g. fast exchange of the protonated and deprotonated forms of histidine (Section 2.2); the absence of certain splittings in the ^{1}H spectrum of

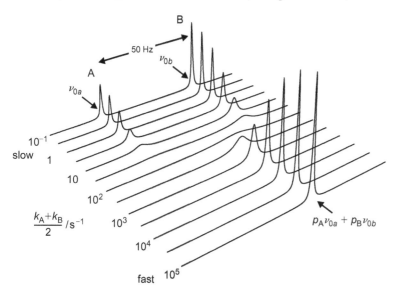

Fig. 4.7 Calculated NMR spectra for a pair of nuclei exchanging between two sites A and B with populations in the ratio $p_B/p_A=2$ (unsymmetrical two-site exchange). Spectra are shown for a range of values of the mean rate constant $\frac{1}{2}(k_A+k_B)$. The difference in resonance frequencies of the two sites, $\delta\nu$, is 50 Hz. The linewidths in the absence of exchange are 1 Hz.

ethanol (Section 3.3); the averaging of three-bond coupling constants by rapid internal rotation (Section 3.7).

4.4 Examples

A number of the following examples involve ^{13}C NMR spectra, recorded with proton decoupling, denoted $^{13}C\{^1H\}$, at natural isotopic abundance of ^{13}C, so that each inequivalent carbon site gives rise to an NMR singlet, in the absence of other magnetic nuclei. $^{13}C\{^1H\}$ NMR is particularly suited to the study of chemical exchange effects, whose theoretical analysis can become quite complicated in the presence of J-couplings.

Cis-decalin–ring inversion

Cis-decalin, $C_{10}H_{18}$, consists of two cyclohexane rings, both in a chair conformation, fused cis to one another. Unlike the trans isomer, which is conformationally rigid, cis-decalin flips between two degenerate conformations by chair-to-chair inversion of both rings, a process clearly revealed by $^{13}C\{^1H\}$ NMR (Fig. 4.8). At

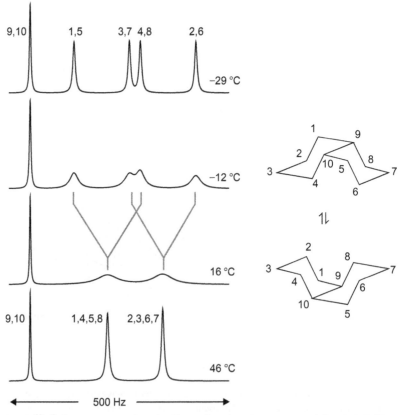

Fig. 4.8 $^{13}C\{^1H\}$ NMR spectra of cis-decalin as a function of temperature. (After D. K. Dalling, D. M. Grant, and L F. Johnson, J. Am. Chem. Soc., **93** (1971) 3678.)

low temperature, when the inversion is slow, equally intense resonances are seen from the five pairs of equivalent carbons: (1,5), (2,6), (3,7), (4,8), and (9,10). Ring inversion at higher temperatures interchanges the chemical shifts as follows: $1 \leftrightarrow 4$, $5 \leftrightarrow 8$, $2 \leftrightarrow 3$, and $6 \leftrightarrow 7$, causing the (1,5) and (4,8) peaks to merge into a single line, and similarly the (2,6) and (3,7) resonances. The fifth ^{13}C signal, from carbons 9 and 10, is not affected, and remains sharp at all temperatures.

This simple example of symmetric two-site exchange may be analysed using the results of Section 4.2. Consider the (1,5) and (4,8) resonances, which have a chemical shift difference of 7.0 ppm, corresponding to a frequency difference δv of 175 Hz at the ^{13}C NMR frequency of 25 MHz. Rate constants for the inversion are easily determined as follows. At $-29\,°C$, the line broadening from slow exchange (Δv) is 2.9 Hz, giving $k = 2.9\pi \approx 9\ s^{-1}$ (eqn 4.2). By $+27\,°C$, the exchange is fast and the two lines have merged to give a single line with a broadening of 24.3 Hz. Using eqn 4.3, we have $k = \frac{1}{2}\pi(175)^2/24.3 \approx 2000\ s^{-1}$. The (1,5) and (4,8) peaks coalesce at around $+5\,°C$, at which point $k = \pi(175)/\sqrt{2} \approx 400\ s^{-1}$ (eqn 4.4). Since the difference in chemical shifts of the two other pairs of exchanging carbons is almost the same as for (1,5) and (4,8), they coalesce at roughly the same temperature and have similar widths for fast exchange.

Bullvalene—degenerate rearrangement

A more complex example of the interconversion of chemically identical forms of a molecule is provided by bullvalene, a molecule comprising ten CH groups in a cage structure with a three-fold axis of symmetry (Fig. 4.9).

At $-63\,°C$ in $CHCl_2$–$CHCl_2$ solution, the 22.5 MHz ^{13}C{^{1}H} spectrum consists of four resonances, from the four distinct carbon sites, with intensity ratios 3:3:3:1, reflecting the symmetry of the molecule. As the temperature is increased, all four broaden and merge into a single sharp peak at the weighted-average chemical shift, $(3\delta_a + 3\delta_b + 3\delta_c + \delta_d)/10$.

The process responsible is a series of degenerate Cope rearrangements, in which the CH groups are permuted without changing the structure. Each molecule may rearrange in three ways by breaking a different cyclopropane bond, leading to a complex network of thousands of interconverting forms, with all possible permutations of CH groups.

A detailed analysis using computer-simulated spectra yields the rate constants shown in Fig. 4.9. An Arrhenius plot ($\ln k$ against $1/T$) gives an activation energy $E_a \approx 60\ kJ\ mol^{-1}$ and pre-exponential factor $A \approx 10^{14}\ s^{-1}$. Similar exchange effects are observed in the ^{1}H spectrum, but are more complicated to interpret because of the ^{1}H–^{1}H J-couplings.

Finally, it is interesting to note that resonance b is narrower in the slow exchange regime than the other three, which all have similar widths (e.g. at $-37\,°C$). The reason is that every time the molecule undergoes a rearrangement, only one of the three carbons in site b changes its Larmor frequency (the other two stay in site b); all seven remaining carbons are moved to a different site (see Fig. 4.9, bottom right). The average lifetime of the carbons in site b is therefore three times longer, and the exchange broadening three times smaller than for sites a, c, and d.

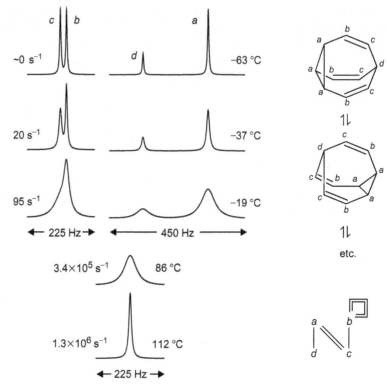

Fig. 4.9 $^{13}C\{^1H\}$ NMR spectra of bullvalene as a function of temperature. Approximate exchange rates are shown to the left of each spectrum. The scheme at bottom right summarizes the interchange of the carbon sites. For example, the bond rearrangement shown on the right moves two of the three carbons in position a to position c and the other one to position d. (After H. Günther and J. Ulmen, *Tetrahedron*, **30** (1974) 3781.)

This differential line broadening can be used to determine the mechanism of a rearrangement, as the following example illustrates.

$(\eta^4\text{-}C_8H_8)Ru(CO)_3$—mechanism of rearrangement

Below about $-100\,°C$, the eight carbons of the cyclooctatetraene ligand in $(\eta^4\text{-}C_8H_8)Ru(CO)_3$ give rise to four equally intense ^{13}C resonances, consistent with a structure in which the ruthenium is attached to two of the four double bonds of the C_8H_8 ring (Fig. 4.10). No ruthenium–carbon couplings are observed because of rapid quadrupolar relaxation (see Section 5.7) of both ^{99}Ru $(I=\frac{5}{2})$ and ^{101}Ru $(I=\frac{5}{2})$, the only magnetic isotopes of this element. As the temperature is increased, peaks b and c broaden more rapidly than a and d. Above about $-60\,°C$, a single ^{13}C resonance is observed at the average chemical shift of the four peaks indicating rapid rearrangement of the ligand, rendering all eight carbons effectively equivalent.

Figure 4.10 shows the four possible $1,n$ shifts ($n=2, 3, 4, 5$) of the $Ru(CO)_3$ group. A 1,5 shift can be excluded immediately: it simply swaps sites a and d, and

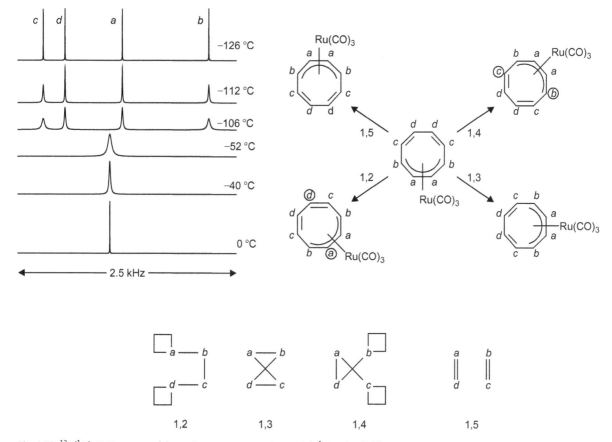

Fig. 4.10 $^{13}C\{^1H\}$ NMR spectra of the cyclooctatetraene carbons of $(\eta^4\text{-}C_8H_8)Ru(CO)_3$ as a function of temperature. The observed pattern of linebroadening is only compatible with a 1,2 rearrangement mechanism. (After F. A. Cotton and D. L. Hunter, *J. Am. Chem. Soc.,* **98** (1976) 1413.) The carbons whose labels are circled do not change their chemical shifts when the $Ru(CO)_3$ group moves around the ring. The schemes at the bottom summarize the interchange of carbon sites by the four possible 1,n rearrangements.

sites b and c, and so would lead to *two* peaks in fast exchange at chemical shifts $\frac{1}{2}(\delta_a + \delta_d)$ and $\frac{1}{2}(\delta_b + \delta_c)$.

Of the remaining rearrangement pathways, only a 1,2 shift is consistent with the low temperature spectra. As shown in Fig. 4.10, a 1,2 jump has no effect on the chemical shift of *one* of the carbons in each of sites a and d, but moves *both* of the b and the c carbons to another site with a different resonance frequency. On average, therefore, sites a and d have *twice* the lifetime of b and c and show *half* the line broadening. A 1,4 shift of the $Ru(CO)_3$ group has exactly the opposite effect on the four sites and would cause peaks a and d to have *twice* the width of b and c. By contrast, a 1,3 jump of the $Ru(CO)_3$ group alters the resonance frequency of all eight carbons and so would broaden all four peaks to the same extent; a random jump mechanism would have a similar effect.

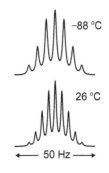

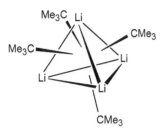

Fig. 4.11 $^{13}C\{^1H\}$ NMR spectra of the t-butyl α carbon of $[^6Li^{13}CMe_3]_4$ in cyclopentane. The septet ($-88\,°C$) and nonet ($+26\,°C$) are formed by coupling to, respectively, three and four equivalent $I=1$ ^6Li nuclei. (After R. D. Thomas, M. T. Clarke, R. M. Jensen, and T. C. Young, *Organometallics*, **5** (1986) 1851.)

[Li(*t*-Bu)]₄–averaging of *J*-couplings

So far the impression has been given that the only NMR properties that can be averaged by fast exchange are chemical shifts. In fact the same principles apply to spin-spin couplings and relaxation times, although we shall not discuss the latter here.

In cyclopentane solution, t-butyl lithium is tetrameric: the lithium atoms are placed at the corners of a tetrahedron with a Me_3C group sitting above each of the triangular faces. Consistent with this, the α carbons of the t-butyl groups in the ^6Li compound at temperatures below $-22\,°C$ show a single $^{13}C\{^1H\}$ resonance with seven equally spaced lines (Figs 4.11 and 3.12). Each ^{13}C couples equally to three equidistant ^6Li nuclei to give a 1:3:6:7:6:3:1 multiplet pattern (^6Li has $I=1$ but undergoes *slow* quadrupolar relaxation), with no resolvable coupling to the fourth, more remote, ^6Li. The $^6Li-^{13}C$ coupling constant is 5.4 Hz.

At temperatures above $-5\,°C$, the septet is replaced by a nonet (nine lines) with splitting 4.1 Hz. Rapid intramolecular scrambling of the t-butyl groups causes the ^{13}C to interact equally with all four lithiums, with a weighted average coupling constant: $\frac{3}{4}\times5.4+\frac{1}{4}\times0.0\approx4.1$ Hz.

A further example of the averaging of coupling constants is provided by liquid ethanol.

Ethanol–proton exchange

In Chapter 3 we saw how the interaction between the CH_3 and CH_2 protons in liquid ethanol leads to a triplet and a quartet in the ^1H spectrum. But what of the other three-bond coupling in the molecule, between the OH and CH_2 protons? The splitting due to this interaction is only visible in highly purified ethanol, from which all traces of acid and base have been removed. Under these stringent conditions, the OH singlet splits into a 1:2:1 triplet ($J=4.8$ Hz), and each line of the CH_2 quartet becomes a doublet with the same coupling constant (Fig. 4.12). On addition of as little as 10^{-5} M acid, these extra splittings vanish as a result of rapid acid-catalysed H^+ exchange.

Consider an ethanol molecule in which the hydroxyl proton (A) exchanges with a H^+ ion (B) as shown at the bottom of Fig. 4.12. If H_A and H_B have the same magnetic quantum number ($m=\pm\frac{1}{2}$), the exchange of protons has no effect on the resonance frequency of the CH_2 protons, and is undetectable by NMR. However, when H_A and H_B have opposite spins ($\uparrow$ and $\downarrow$), the net effect of the exchange is

$$CH_3-CH_2-OH_A\uparrow + H_B^+\downarrow \rightleftharpoons CH_3-CH_2-OH_B\downarrow + H_A^+\uparrow.$$

The ethyl protons cannot tell whether the hydroxyl proton is A or B, only whether it is $\uparrow$ or $\downarrow$. So, at each H^+ exchange, the CH_2 resonance frequency changes by $\pm J$. If this hopping is fast compared to the frequency difference involved, i.e. J ($=4.8$ Hz), the doublet splitting of the CH_2 multiplet collapses to zero, in the same way that chemical shift differences are removed by fast exchange. The triplet structure of the OH resonance also disappears when proton exchange is fast compared to J.

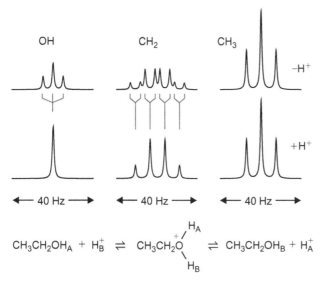

Fig. 4.12 ^{1}H NMR multiplets of ethanol with (below) and without (above) a trace of acid to catalyse intermolecular hydrogen transfer.

The influence of H^+ exchange is also evident in the ^{1}H spectrum of ethanol-water mixtures (Fig. 4.13). When acid is rigorously excluded, separate OH resonances for ethanol and water can be seen; addition of a little acid leads to fast exchange and a single peak at the average chemical shift, weighted by the relative concentrations of H_2O and ethanol OH protons.

Finally, there is one more exchange process in ethanol which has rather been taken for granted so far. This is the averaging, by rapid internal rotation around the C–C and C–O bonds, of the chemical shift differences of the three methyl protons, and of the two methylene protons, which arise from the various conformations adopted by the molecule. At sufficiently low temperatures, when the rate of internal rotation is comparable to these frequency differences, the ^{1}H spectrum of this endlessly fascinating molecule would therefore become more complicated than Fig. 4.12.

Motion of tyrosine side-chains in proteins

Advances in NMR techniques have made it possible to determine the solution-phase structures of proteins of up to about 25,000 relative molecular mass (Section 5.8). Important though it is to know their structures, proteins are far from rigid and information on their dynamical properties is essential for a complete understanding of their functions. A simple way in which NMR can provide such insights is through the exchange effects associated with the internal rotation of groups such as the phenol side-chain of the amino acid tyrosine.

In a rigid protein, the four ring protons of a tyrosine have different local environments and, barring coincidences, different chemical shifts. In particular, the

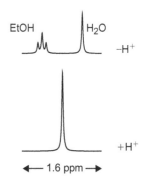

Fig. 4.13 ^{1}H spectra of the OH resonances of ethanol and water with (below) and without (above) a trace of acid.

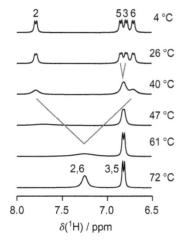

8.0   7.5   7.0   6.5

$\delta(^1\text{H})$ / ppm

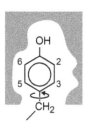

Fig. 4.14 ^{1}H NMR signals of Tyr-35 in basic pancreatic trypsin inhibitor, BPTI, as a function of temperature. The structure indicates the asymmetric environment of a tyrosine side-chain in a protein. The ring flipping rates are $2.5\times10^{-2}\text{ s}^{-1}$ at 4 °C, and $1.7\times10^{4}\text{ s}^{-1}$ at 72 °C. (After K. Wüthrich, *NMR of Proteins and Nucleic Acids*, Wiley, New York, 1986.)

Answers to the exercises are provided at the back of the book. Full worked solutions are available on the Online Resource Centre at www. oxfordtextbooks.co.uk/orc/hore_nmr2e/

asymmetric surroundings (Fig. 4.14) cause H2 and H6 to be inequivalent, and similarly H3 and H5. However, if the ring undergoes 180° flips around its two-fold axis, H2 and H6 swap chemical shifts, as do H3 and H5, and if this exchange is fast compared to the frequency differences, only two ^{1}H resonances will be observed, at positions $\frac{1}{2}(\delta_2+\delta_6)$ and $\frac{1}{2}(\delta_3+\delta_5)$.

The ring protons of a rigid tyrosine residue in a protein typically cover a chemical shift range of 1 ppm, so that on a 500 MHz spectrometer lineshape changes are detectable for flipping rates between a few times per second up to about 10^5 s^{-1}. The rate at which an aromatic ring rotates is governed principally by the dynamics of neighbouring groups in the protein which must move aside to allow the ring to flip. The ^{1}H NMR spectra of tyrosine and phenylalanine (tyrosine without the hydroxyl group) side-chains thus give information on these relatively slow cooperative structural fluctuations, over a broad timescale.

As an example, Fig. 4.14 shows a series of ^{1}H spectra of basic pancreatic trypsin inhibitor. BPTI is a small globular protein of relative molecular mass 6500, consisting of a single polypeptide chain of 58 amino acid residues, of which four are tyrosines. Below about 30 °C, Tyr-35 exhibits four separate aromatic resonances, which merge into two above 70 °C. Similar behaviour is found for Tyr-23 (not shown), at a temperature some 30 °C lower. Both Tyr-10 and Tyr-21 give two resonances, even at 4 °C, and are essentially free rotors.

4.5 Summary

- Chemical exchange effects in NMR arise from dynamic chemical and conformational equilibria.

- NMR lines are broadened by slow exchange.

- Differences in the NMR frequencies of exchanging spins (δv) can be averaged by fast exchange.

- The magnitude of δv relative to the exchange rate constant(s) determines whether the exchange is 'slow' or 'fast'.

- Chemical exchange effects give information on the rates and mechanisms of chemical reactions, molecular rearrangements, and internal motions.

4.6 Exercises

1. Two protons in a molecule in a 400 MHz spectrometer undergo symmetrical two-site exchange. Their chemical shifts in the absence of exchange are 2.0 and 6.0 ppm. Determine the line broadening resulting from the exchange process when the rate constant is (a) 10^2 s^{-1} and (b) 10^5 s^{-1}. The spectrometer frequency is 400 MHz.

2. A molecule undergoes symmetrical two-site chemical exchange. In the slow exchange regime, the additional ^{1}H linewidths arising from the

exchange are 1.4 Hz at 100 °C and 6.0 Hz at 120 °C. Determine (a) the rate constants (k) at the two temperatures and hence (b) the activation energy (E_a) and the Arrhenius pre-exponential factor (A) for the equilibrium. NB $k = A\exp(-E_a/RT)$.

3. The rate constant for a symmetrical two-site exchange has the temperature dependence: $k/(s^{-1}) = 10^{13}\exp[-9100/(T/K)]$. The ^1H chemical shift difference between the two sites is 1.0 ppm. Determine the appearance of the 400 MHz spectra at the following temperatures: (a) 310 K; (b) 393 K; (c) 420 K.

4. Two protons in a molecule undergo unsymmetrical two-site conformational exchange. Their chemical shifts in the absence of exchange are 3.5 and 6.5 ppm. Under conditions of fast exchange, the observed chemical shift is 3.86 ppm at 300 K and 4.10 ppm at 350 K. Determine the difference in molar enthalpy between the two conformers.

5. A molecule undergoes unsymmetrical two-site chemical exchange, $A \rightleftharpoons B$. When the exchange is slow, the ^1H spectrum contains two lines with chemical shifts $\delta_A = 3.00$ ppm and $\delta_B = 5.00$ ppm. In fast exchange a single line is observed at 3.01 ppm. What are the fractional populations of the two forms?

6. The unsymmetrical two-site exchange $A \rightleftharpoons B$ has an equilibrium constant $K = 0.01$. Under slow exchange conditions, the linewidth of a nucleus in species A is 1.0 Hz. What is the linewidth of the corresponding nucleus in species B? Assume that the linewidths in the absence of exchange are negligible.

7. Addition of 10^{-7} M HCl to a solution of pure methanol in water produces a linebroadening of the OH proton signal of 3.2 Hz. What is the second-order rate constant for the H^+ exchange reaction?

8. At temperatures above 100 °C, [18]-annulene (Fig. 2.25(c)) undergoes a rapid internal rearrangement in which the inner and outer hydrogen atoms are exchanged. Predict the chemical shift of the single ^1H line observed at 100 °C.

9. The aromatic region of the $^{13}C\{^1H\}$ spectrum of N-methylaniline contains six lines at low temperature. How many lines might be expected at high temperature?

10. The pK_a of $H_2PO_4^-$ is 7.21 and the ^{31}P chemical shifts of $H_2PO_4^-$ and HPO_4^{2-} are 3.40 and 5.82 ppm, respectively. Under conditions of fast exchange $(H_2PO_4^- \rightleftharpoons HPO_4^{2-} + H^+)$ the ^{31}P spectrum contains a single line at 4.61 ppm. What is the pH of the solution?

5 Spin relaxation

5.1 Introduction

Chemical shifts, spin–spin couplings, and chemical exchange are the factors that chiefly determine the appearance of liquid-state NMR spectra. More subtle, but no less important, are the two relaxation processes—*spin–lattice relaxation* and *spin–spin relaxation*—which cause nuclear spins to return to equilibrium following some disturbance.

The sign of the Larmor frequency is irrelevant for the material in this chapter. To avoid extensive use of modulus signs ($|\omega_0|$), we will simply ignore the fact that ω_0 is negative for a nucleus with $\gamma > 0$. The dependence of Larmor frequencies on chemical shifts is also of minimal importance for spin relaxation. We therefore use $\omega_0 = |\gamma| B_0$ throughout.

5.2 Spin–lattice relaxation

Imagine doing an NMR experiment. Before the sample is put into the magnet, the two energy levels of a spin-$\frac{1}{2}$ nucleus are degenerate (neglecting the tiny effect of the Earth's magnetic field) and their populations are equal. As soon as the sample is in the magnet, the energy levels split apart and if nothing else happened the populations would remain equal and NMR would be impossible. Fortunately, *spin–lattice relaxation* comes to the rescue by enabling the spins to move between their energy levels so as to establish the Boltzmann population difference (eqn 1.11). As the populations approach equilibrium, the energy released is dissipated in the surroundings (traditionally known as the 'lattice').

These changes in populations are characterized by a time T_1—*the spin-lattice relaxation time*. For a collection of spin-$\frac{1}{2}$ nuclei, assuming the relaxation is exponential (which is often the case), the difference in the numbers of $m = \pm\frac{1}{2}$ spins grows according to

$$\Delta n(t) = \Delta n_{eq}\left[1 - \exp(-t/T_1)\right], \tag{5.1}$$

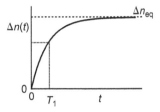

Fig. 5.1 The difference ($\Delta n(t)$) in $m = \pm\frac{1}{2}$ populations of a collection of spin-$\frac{1}{2}$ nuclei as a function of the time after putting the sample into a magnetic field, eqn 5.1. When $t = T_1$, the population difference has grown to 63% of its value (Δn_{eq}) at thermal equilibrium.

where Δn_{eq} is the equilibrium population difference and t is time (see Fig. 5.1). After putting the sample in the magnet, one should therefore wait a time long compared to T_1 before trying to record a spectrum. This is not normally a problem as typical T_1 values for spin-$\frac{1}{2}$ nuclei in liquids are usually no more than a few seconds. As we shall see in Section 6.4, there are much better ways of measuring T_1 values than simply dropping the sample into a magnet and trying to watch the NMR signal appear.

Similar behaviour can be expected whenever non-equilibrium spin populations are produced. Equation 5.1 is a particular solution of the general rate equation

$$\frac{\mathrm{d}\Delta n}{\mathrm{d}t} = -\frac{1}{T_1}(\Delta n - \Delta n_{\mathrm{eq}}). \tag{5.2}$$

The reason why relaxation restores the equilibrium population difference, Δn_{eq}, is fundamentally that the rate of downward spin transitions $(\beta \to \alpha)$ exceeds the rate of upward spin transitions $(\alpha \to \beta)$ by a factor $\exp(\hbar\omega_0/k_{\mathrm{B}}T)$.

The origin of spin–lattice relaxation

Relaxation mechanisms that are important in other forms of spectroscopy are generally ineffective in NMR. The rate of spontaneous emission (fluorescence), for example, scales with the cube of the transition frequency and is negligibly slow at NMR frequencies. Nuclear spins interact so weakly with the rest of the world that they are almost completely decoupled from the motions of the molecules that contain them. As a molecule rotates, its nuclear spins remain aligned with the magnetic field, rather than reorienting with the molecule; in this respect a spin behaves like a ship's gimbal compass which points north however much the ship rolls, pitches, and yaws.

The mechanism of nuclear spin relaxation lies, not surprisingly, in magnetic interactions, the most important being dipolar coupling. As described in Appendix A and Section 3.8, the dipolar coupling between two nuclei depends on their separation r and on θ, the angle between the internuclear vector and the magnetic field direction. Although this purely anisotropic coupling does not normally lead to splittings in liquid-phase NMR spectra, the *instantaneous* interaction is far from negligible. As the molecules translate, rotate, and vibrate in solution, r and θ vary in a complicated way causing the interaction to fluctuate rapidly. Thus the dipolar coupling, modulated by molecular motions, causes nuclear spins to experience time-dependent local magnetic fields which, if they contain a component at the Larmor frequency, can induce radiationless transitions that return the spins to equilibrium. Most other spin relaxation mechanisms have essentially the same origin: an intramolecular or (less commonly) an intermolecular magnetic interaction (or electric quadrupole interaction for $I > \frac{1}{2}$ nuclei), rendered time-dependent by random molecular motion.

Before saying more about spin–lattice relaxation, we need to look a little at the rotational motion of molecules in liquids.

5.3 Rotational motion in liquids

In gases, at least at low pressure, the mean free path is large and a molecule can rotate end-over-end many times before suffering a collision that changes its rotational state. In liquids, collisions occur much more frequently; molecules are buffeted constantly from all sides, each bump accelerating or decelerating the rotational motion and deflecting the rotation axis. As noted before,

Fig. 5.2 A typical random walk trajectory for a molecule undergoing rotational diffusion in a liquid.

this motion is often referred to as *tumbling* rather than rotation, to reflect its chaotic nature.

Let us focus on a simple molecule, CH_4, and imagine the carbon atom fixed at the centre of a sphere with radius equal to the C–H bond length. As the methane molecule collides with its neighbours in solution, each of the four hydrogen atoms undergoes a random walk on the surface of the sphere. Figure 5.2 shows one such zigzag trajectory for an H atom, initially at the 'north pole'. Every CH_4 undergoes a different random walk, determined by its collisions with other molecules. We can visualize this by plotting, for many such trajectories, the position reached by the H atom a certain time after setting off from the north pole (Fig. 5.3). After a short time (a), the molecules have spread out a bit on the sphere, but are still clustered around their starting position. Somewhat later (b), the average displacement has increased and some have reached the equator. After a longer time, there are substantial numbers of H atoms in the southern hemisphere but still an excess at northern latitudes (c). Only if we were to wait a very long time would the molecular orientations be randomly distributed around the sphere.

We can define a characteristic time for this motion, the *rotational correlation time*, τ_c. Roughly speaking, τ_c is the time taken for the root-mean-square deflection of the molecules to be about 1 radian ($\approx 57°$). At times much less than τ_c, most of the molecules are close to their original positions and when $t \gg \tau_c$ the orientations are completely randomized and all 'memory' of the original orientation is lost. Typical values for τ_c for small molecules in non-viscous solvents at room temperature are in the region of 100 ps.

The point of Figs 5.2 and 5.3 is that they allow us to see, qualitatively, what the frequency spectrum of this random motion ought to look like. Since τ_c is the average time taken to tumble through about a radian, τ_c^{-1} is approximately the root-mean-square rotational frequency (in rad s^{-1}). Further, Fig. 5.3 shows that frequencies *less than* τ_c^{-1} are quite probable, because they correspond to rotations of less than a radian during a time τ_c, while frequencies *above* τ_c^{-1}, which correspond to rotations greater than a radian, will be much less likely.

In short, the frequency spectrum of an intramolecular magnetic interaction, modulated by molecular tumbling, should resemble one of the curves drawn in

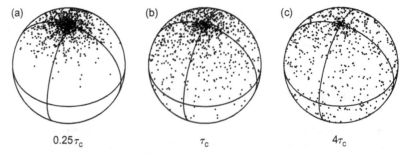

Fig. 5.3 Representative positions of molecules undergoing rotational diffusion in a liquid, at different times after starting off from the 'north pole': (a) $t = \frac{1}{4}\tau_c$; (b) $t = \tau_c$; (c) $t = 4\tau_c$, where τ_c is the rotational correlation time.

Fig. 5.4. This function is given the symbol $J(\omega)$ and is called the *spectral density*; ω is the angular frequency (in rad s^{-1}). $J(\omega)$ can be thought of as being proportional to the *probability* of finding a component of the random motion at frequency ω. Smaller molecules, or less viscous solvents, or higher temperatures should all result in shorter correlation times (faster tumbling, on average) and hence a spectral density that extends to higher frequencies, as shown in Fig. 5.4. These curves have been drawn using the most common form of $J(\omega)$:

$$J(\omega) = \frac{2\tau_c}{1 + \omega^2 \tau_c^2}, \tag{5.3}$$

which is appropriate when the molecule's 'memory' of its orientation at an earlier time decays *exponentially*.

5.4 Spin–lattice relaxation again

Although the dipolar interaction is the most common source of relaxation, it is not the simplest. Two dipolar-coupled spins experience *correlated* time-dependent magnetic fields (they have the same r and θ in eqns 3.15 and 3.16) and consequently relax in a concerted manner. Although this is the source of some interesting and useful relaxation effects (Section 5.5), it is an unwelcome complication at this stage. To keep things simple, the following paragraphs discuss an idealized mechanism in which nuclear spins are independently relaxed by random local fields (Slichter 1990). This approximation gives insight into spin relaxation in general, and requires relatively minor adjustments to yield quantitative predictions for particular relaxation mechanisms.

Spin–lattice relaxation results from fluctuating local fields that induce nuclei to undergo radiationless transitions between their spin energy-levels. The rate constant for this process, T_1^{-1}, depends on the probability that the local fields have a component that oscillates at the Larmor frequency, ω_0. In other words T_1^{-1} is proportional to the spectral density at the Larmor frequency, $J(\omega_0)$. The predicted relaxation rate is

$$\frac{1}{T_1} = \gamma^2 \left\langle B_{loc}^2 \right\rangle J(\omega_0), \tag{5.4}$$

where $\left\langle B_{loc}^2 \right\rangle$ is the mean-square value of the local fluctuating fields and γ is the magnetogyric ratio of the spin in question. Not surprisingly, stronger local fields lead to faster relaxation, other things being equal.

We now have the information necessary to predict how the spin–lattice relaxation rate varies with the rotational correlation time (Fig. 5.5). For rapidly tumbling molecules with $\omega_0 \tau_c \ll 1$ (left-hand side of Fig. 5.5), $J(\omega_0) \approx 2\tau_c$ and $T_1 \propto 1/\tau_c$ so that the relaxation becomes *faster* as τ_c is increased (e.g. by lowering the temperature). Conversely, slowly tumbling molecules have $\omega_0 \tau_c \gg 1$ (right-hand side of Fig. 5.5), $J(\omega_0) \approx 2/\omega_0^2 \tau_c$ and $T_1 \propto \tau_c$, so that the relaxation becomes *slower* as τ_c is increased. As τ_c is varied, T_1 goes through a minimum (maximum relaxation rate) when $\tau_c = 1/\omega_0$ as may be seen by differentiation of eqn 5.3.

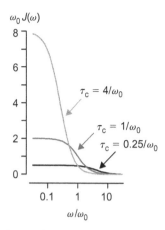

$\omega_0 J(\omega)$

ω/ω_0

Fig. 5.4 The spectral density function $J(\omega)$, scaled by the Larmor frequency ω_0, drawn for three values of the rotational correlation time τ_c. The horizontal axis is logarithmic. As τ_c is reduced, $J(\omega)$ extends to higher frequencies and is generally flatter.

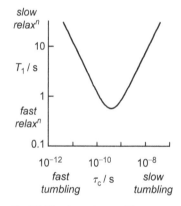

Fig. 5.5 The dependence of the spin–lattice relaxation time T_1 on the rotational correlation time τ_c, from eqns 5.3 and 5.4, when $\omega_0/2\pi = 400$ MHz. Both axes are logarithmic. The regions of the graph corresponding to fast and slow tumbling, and fast and slow relaxation are indicated. The value of $\gamma^2 \left\langle B_{loc}^2 \right\rangle$ used (4.5×10^9 rad^2 s^{-2}) is roughly appropriate for the dipolar coupling of two protons separated by 2 Å. The minimum T_1 occurs when $\tau_c = 1/\omega_0 \approx 400$ ps. On a higher field spectrometer, the minimum would come at a smaller value of τ_c.

To decide whether a given molecule falls on the left or the right of the T_1 minimum (Fig. 5.5), one clearly needs to know τ_c. A rough and ready rule of thumb (Sanders and Hunter 1993) is that for molecules in water at room temperature, τ_c in picoseconds is very approximately equal to the relative molecular mass M_r. For example, if $M_r = 100$, $\tau_c \approx 100$ ps; if $M_r = 10{,}000$, $\tau_c \approx 10$ ns. On a 400 MHz spectrometer, the maximum ^{1}H relaxation rate occurs when $\tau_c \approx 1/\omega_0 \approx 400$ ps, so that most molecules with an M_r of a couple of hundred or less should fall to the left of the T_1 minimum, while those with a molecular weight of a thousand or more come on the right. Molecules are expected to have longer correlation times in more viscous solvents and at lower temperatures.

It can now be seen why *rotational* motion is important for most spin relaxation mechanisms. *Vibrations* are usually much too fast to have a significant component at the relatively low NMR frequencies (ω_0). Modulation of *inter*molecular dipolar interactions by *translational* motion is also relatively inefficient because the couplings are generally weaker (larger average distance between spins) than in the intramolecular case. *Rotation*, by contrast, occurs at roughly the right frequency and modulates *intra*molecular interactions; it therefore tends to be the dominant motion for spin relaxation.

From Appendix A, and in particular eqn A.1, we can see that the dipolar field $(\sim \langle B_{loc}^2 \rangle^{1/2})$ depends on r^{-3} so that the spin–lattice relaxation rate of a pair of dipolar-coupled spins should be proportional to r^{-6}. Spin–lattice relaxation times are thus sensitive functions of internuclear separations and hence molecular structure.

The predictions of this rough and ready argument are confirmed by a more detailed theoretical treatment which gives the following expression for the spin–lattice relaxation rate of two dipolar-coupled nuclei, A and X, with different magnetogyric ratios, in the so-called *extreme narrowing limit*, $\omega_0 \tau_c \ll 1$:

$$\frac{1}{T_1} = (2\pi R_{AX})^2 \tau_c = \left(\frac{\mu_0}{4\pi}\right)^2 \hbar^2 \gamma_A^2 \gamma_X^2 \left(\frac{\tau_c}{r^6}\right) \tag{5.5}$$

in which τ_c comes from the spectral density (eqns 5.3 and 5.4) and R_{AX} is the dipolar coupling constant (eqn 3.15). Typical dipolar T_1 values for protons are in the range 0.1–10 s (see also Fig. 5.5). Nuclear spin relaxation is slow for two reasons. The local magnetic fields are generally rather feeble and interact weakly with nuclear spins, and even when $\omega_0 \tau_c = 1$ the spectral density at the Larmor frequency is small. The dependence of T_1^{-1} on the magnetogyric ratios means that, other things being equal, ^{13}C nuclei should relax more slowly than protons.

There are two further points about spin–lattice relaxation. First, why do the transitions induced by local time-dependent fields return the spins to equilibrium? Second, when the spins relax, energy is either released or absorbed; where does this energy go to or come from?

Spin–lattice relaxation couples the spins (very weakly) to the motion of the molecules that carry them and so provides a pathway for the exchange of energy between the spin system and its surroundings. In other words, spin–lattice relaxation brings the spins into thermal contact with the lattice, enabling them to come to equilibrium with the rest of the world. As mentioned in Section 5.2, this restores the equilibrium populations because the rate of downward spin relaxation $(\beta \to \alpha)$ is faster than the upward rate $(\alpha \to \beta)$ by the factor $\exp(\Delta E/k_B T)$, where $\Delta E = \hbar\omega_0$. The energy absorbed or released in the course of spin relaxation is transferred from or to the motions of the molecules, causing a slight cooling or warming of the lattice. Since the spin energies are minute compared to the rotational, vibrational, and translational energies of molecules in solution, nuclear spins are relaxed with an unmeasurably small change in the temperature of the sample, just as dropping a small hot object into a large lake would have a negligible effect on the overall temperature of the water.

Some illustrations and applications of spin–lattice relaxation are presented in Sections 5.8 and 5.9; but first we look at an important relaxation phenomenon resulting from the dipolar relaxation mechanism.

5.5 The nuclear Overhauser effect

Consider a molecule containing two inequivalent protons, A and X, with no J-coupling, so that the ^{1}H spectrum consists of a singlet at each of the chemical shifts. Suppose that, while recording the spectrum, the X-spins are saturated (i.e. their α and β populations are equalized) by applying a strong radiofrequency field at the resonance frequency of X. This destroys the NMR signal of X, but it can also affect the intensity of the A-resonance if the two spins have an appreciable *dipolar* interaction. As shown schematically in Fig. 5.6, the A-peak may get stronger, or weaker, or even invert. This remarkable phenomenon is known as the *nuclear Overhauser effect* (or *enhancement*), NOE. As we shall see, it gives information on internuclear separations much more directly than does spin–lattice relaxation.

To understand the origin of the NOE, it is necessary to look at the possible spin–lattice relaxation pathways available to a pair of dipolar-coupled protons. Figure 5.7 shows the four energy levels: $\alpha_A\alpha_X$, $\alpha_A\beta_X$, $\beta_A\alpha_X$, $\beta_A\beta_X$. To reduce clutter, we henceforth omit the A and X subscripts. Ignoring the chemical shift difference and using $\hbar\omega_0 \ll k_B T$, the relative populations of the four states at equilibrium are

$$n(\alpha\alpha) = 1+2\Delta\,;\; n(\alpha\beta) = n(\beta\alpha) = 1;\; n(\beta\beta) = 1-2\Delta$$

where $\Delta = \frac{1}{2}\hbar\omega_0/k_B T$. The population differences are 2Δ for all four allowed NMR transitions ($\Delta m_A = \pm 1$ or $\Delta m_X = \pm 1$).

Of the six relaxation pathways open to the two coupled spins, four correspond to a *single* spin flipping, i.e. $\alpha_A \leftrightarrow \beta_A$ or $\alpha_X \leftrightarrow \beta_X$, and are nothing more than the spin–lattice relaxation processes we have just been discussing. Their rate constants are denoted W_1^A and W_1^X, where the subscript indicates that the magnetic quantum number changes by ± 1, and the superscript identifies the spin that relaxes.

The other two relaxation pathways are *cross relaxation* processes in which A and X relax *together*, i.e. $\alpha\alpha \leftrightarrow \beta\beta$ (both spins flipping in the same direction, with rate constant W_2^{AX}) and $\alpha\beta \leftrightarrow \beta\alpha$ (A and X flipping in opposite directions, with

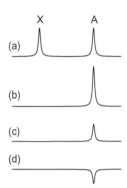

Fig. 5.6 Schematic NMR spectra showing various possible nuclear Overhauser effects. (a) Conventional spectrum of two neighbouring spins A and X. (b)–(d) Spectra resulting from saturation of the X-resonance: the NMR signal of A either gets stronger (b), weaker (c), or inverts (d), depending on the rate of molecular tumbling.

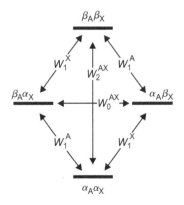

Fig. 5.7 Energy levels for a homonuclear pair of spin-$\frac{1}{2}$ nuclei A and X, showing the six possible relaxation pathways.

rate constant W_0^{AX}). Cross relaxation comes about because the chaotic molecular motion, combined with the mutual dipolar interaction, causes the fluctuating local fields experienced by A and X to be *correlated*: $(3\cos^2\theta - 1)/r^3$ has the same instantaneous value for both spins. The result is that the nuclei can undergo *simultaneous* spin–flips. The W_0^{AX} and W_2^{AX} processes are simply extra pathways that allow the populations to return to equilibrium. Note that spin relaxation does not involve absorption or emission of electromagnetic radiation, and so is not subject to the usual $\Delta m = \pm 1$ selection rule.

Now, to see how the NOE arises, imagine applying a radiofrequency field to spin X of sufficient strength to saturate both X transitions ($\alpha\alpha \leftrightarrow \alpha\beta$ and $\beta\alpha \leftrightarrow \beta\beta$), i.e. to equalize the populations of $\alpha\alpha$ and $\alpha\beta$, and of $\beta\alpha$ and $\beta\beta$ (Fig. 5.8(a) and (b)). We assume, for simplicity, that this can be done without affecting the population differences across the A-transitions ($\alpha\alpha \leftrightarrow \beta\alpha$ and $\alpha\beta \leftrightarrow \beta\beta$), which are still 2Δ. Now we switch off the radiofrequency field and suppose that all relaxation pathways are insignificant except W_2^{AX} (an unrealistic but convenient fiction). This relaxation route transfers population between $\alpha\alpha$ and $\beta\beta$ and ultimately restores their equilibrium populations, $1 + 2\Delta$ and $1 - 2\Delta$, respectively (Fig. 5.8(c)). The population differences across the A-transitions are now 3Δ; that is, the intensity of the A-peak has *increased* (by 50%). Cross relaxation has transferred magnetization from the saturated spin X to its dipolar-coupled partner A.

> In reality, the NOE on A builds up while X is being saturated.

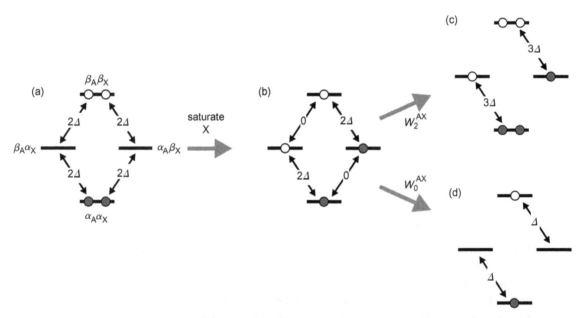

Fig. 5.8 Spin state populations for a homonuclear pair of neighbouring spin-$\frac{1}{2}$ nuclei A and X. Filled/open circles indicate a population excess/deficit of $\Delta = \frac{1}{2}\hbar\omega_0/k_BT$. The arrows between the energy levels are labelled with the appropriate population differences. (a) Thermal equilibrium. (b) Effect of saturating both transitions of spin X without affecting A. (c) and (d) Effect of, respectively, rapid $\alpha\alpha \leftrightarrow \beta\beta$ and $\alpha\beta \leftrightarrow \beta\alpha$ cross relaxation on the populations shown in (b). (c) and (d) have been drawn assuming that W_2^{AX} and W_0^{AX}, respectively, are the *only* operative relaxation processes.

Conversely, if after saturating X the W_0^{AX} process is dominant, the populations of $\alpha\beta$ and $\beta\alpha$ are restored to their equilibrium values (both unity) giving a population difference of only Δ across the A-transitions, i.e. a (50%) *reduction* in the intensity of the A-resonance (Fig. 5.8(d)).

The NOE can be quantified by a parameter η, defined in terms of the perturbed NMR intensity of spin A (i), and its normal intensity (i_0):

$$\eta = \frac{i - i_0}{i_0}. \qquad (5.6)$$

The rough and ready argument above suggests $-\frac{1}{2} \leq \eta \leq +\frac{1}{2}$. An exact treatment shows that the maximum homonuclear NOE is indeed $+\frac{1}{2}$, but that the minimum is -1. In reality, neither W_2^{AX} nor W_0^{AX} dominates the other relaxation pathways and η falls between these two extremes. It can be seen from the above discussion that η has the same sign as $W_2^{AX} - W_0^{AX}$.

But what determines the size and sign of $W_2^{AX} - W_0^{AX}$? We saw in Section 5.4 that the rate of spin–lattice relaxation is proportional to the spectral density at the Larmor frequency ω_0 (because the α and β energy levels are separated by $\hbar\omega_0$). Since $\alpha\alpha$ and $\beta\beta$ are separated by $2\hbar\omega_0$, and $\alpha\beta$ and $\beta\alpha$ are very nearly degenerate, we can expect

$$W_2^{AX} \propto J(2\omega_0) \quad \text{and} \quad W_0^{AX} \propto J(0). \qquad (5.7)$$

Consider first a slowly tumbling molecule with $\omega_0\tau_c \gg 1$. As shown in Fig. 5.9(a), $J(0) \gg J(2\omega_0)$ so that $W_2^{AX} - W_0^{AX} < 0$ and $\eta < 0$. In other words, saturating a proton in a large molecule should cause a *reduction* in the NMR intensity of nearby protons.

For a small molecule, the situation is not quite so clear. If $\omega_0\tau_c \ll 1$, then $J(0) \approx J(2\omega_0)$ (Fig. 5.9(a)). Therefore in the extreme narrowing limit the difference between the two cross relaxation rates arises not from spectral densities but

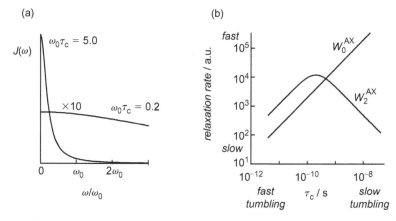

Fig. 5.9 (a) Spectral densities $J(\omega)$ as a function of ω for $\omega_0\tau_c = 5$ (slow tumbling), and $\omega_0\tau_c = \frac{1}{5}$ (fast tumbling). The latter is plotted on an expanded vertical scale. Unlike Fig. 5.4, the ω-axis is linear. (b) The dependence of the two cross relaxation rates W_0^{AX} and W_2^{AX} on τ_c calculated for $\omega_0/2\pi = 400$ MHz. Both axes are logarithmic. The units for the vertical axis are arbitrary. The maximum W_2^{AX} occurs when $\omega_0\tau_c = \frac{1}{2}$.

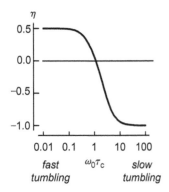

Fig. 5.10 Dependence of the nuclear Overhauser enhancement η (eqn 5.6) on $\omega_0\tau_c$ for an isolated pair of homonuclear spins-$\frac{1}{2}$. The figure shows the maximum homonuclear NOE, observable in the absence of relaxation mechanisms other than the dipolar mechanism. The horizontal axis is logarithmic. $\eta = 0$ when $\omega_0\tau_c \approx 1$.

from the efficiency with which the dipolar interaction connects $\alpha\alpha$ to $\beta\beta$ and $\alpha\beta$ to $\beta\alpha$. It turns out that when $\omega_0\tau_c \ll 1$, $W_2^{AX} = 6W_0^{AX}$ so that $\eta > 0$ (Fig. 5.9(b)). Thus, saturating a proton in a small molecule should boost the NMR intensity of neighbouring protons.

To summarize, the ^1H-^1H NOE, η, should be positive for fast motion and negative for slow motion. The change of sign occurs when $W_0^{AX} = W_2^{AX}$, at which point the effects of the two cross relaxation pathways cancel; this happens when $\omega_0\tau_c \approx 1$ (Fig. 5.10).

So far we have restricted attention to the dipolar relaxation mechanism for the very good reason that it is the most common source of cross relaxation. In fact, most other relaxation mechanisms tend to diminish NOEs by returning population differences to equilibrium via the W_1 pathways, short-circuiting the cross relaxation routes that are crucial for the NOE. The extreme values of the homonuclear NOE in the limits of very fast and very slow motion are therefore only to be expected when the dipolar interaction is the dominant source of relaxation.

NOEs are also observable for heteronuclear pairs of spins; for example, saturation of proton resonances produces a welcome enhancement in the NMR signals of nearby ^{13}C nuclei in small molecules. In the extreme narrowing limit, the enhancement, in the absence of other relaxation mechanisms, is

$$\eta = \frac{1}{2}\frac{\gamma_X}{\gamma_A}, \tag{5.8}$$

where γ_X and γ_A are the magnetogyric ratios of the saturated nucleus (X) and the observed nucleus (A). Thus for X = ^1H and A = ^{13}C, $\gamma_H/\gamma_C \approx 4$ and the maximum NOE is roughly 2; this should be compared to $\eta = \frac{1}{2}$ in the homonuclear case described above.

As Sections 5.8 and 5.9 will reveal, the NOE is exceedingly useful as a source of information on molecular structure. Although the r^{-6} dependence of the cross relaxation rates (by analogy with eqn 5.5) in principle gives internuclear distances directly, in practice matters are a little more involved. Estimates of *relative* separations can often be obtained if the pairs of nuclei concerned are undergoing similar motions *and* if they have comparable contributions from other, non-dipolar relaxation mechanisms. NOEs are also often used semi-quantitatively to decide whether two nuclei are close to one another or not. Despite these limitations, the NOE is still an invaluable source of structural data.

Finally it should be said that selective saturation is certainly not the only technique for measuring NOEs. A more satisfactory way of disturbing spin populations and observing magnetization transfer to nearby nuclei is outlined in Section 6.6. For more on the NOE, see Neuhaus and Williamson (1989) and Hore, Jones, and Wimperis (2015).

5.6 Spin–spin relaxation

In Chapter 4, we saw that slow chemical exchange reduces the lifetime of spin states and causes line broadening. Spin–lattice relaxation should have the same effect, and a broadening of the order of $1/\pi T_1$ (see eqn 4.2) can be expected. However, this is not the only way relaxation processes affect NMR linewidths, and so it is useful to define a new parameter, the *spin–spin relaxation time, T_2*

$$\frac{1}{\pi T_2} = \Delta v, \tag{5.9}$$

where Δv is the relaxation-induced linewidth (instrumental contributions are mentioned in Section 6.2). T_2 has a more fundamental interpretation (Section 6.3), but for now we regard it as a linewidth parameter.

The second contribution to the linewidth, and hence T_2, may be understood from the following argument. Imagine an immobilized pair of dipolar-coupled nuclei in a disordered molecular solid. As described in Section 3.8, pairs of spins with different orientations have different values of θ and therefore different dipolar splittings, $\sim R_{AX}(3\cos^2\theta - 1)$ (eqn 3.16). The result is the Pake pattern shown in Fig. 3.28. Now imagine the same molecules in a liquid. If the molecular tumbling is extremely slow the spectrum will be essentially the same as that of the disordered solid. However, when the tumbling rate τ_c^{-1} becomes comparable to R_{AX}, the modulation of $3\cos^2\theta - 1$ will tend to average the dipolar splitting. This behaviour can be compared to chemical exchange (Section 4.2). If the molecule jumps between two orientations rapidly compared to the difference in the dipolar couplings at those orientations, then the difference will be averaged and a mean dipolar coupling should be observed. Of course, molecules in liquids jump between all possible orientations so that we need to average the dipolar coupling over θ to determine the appearance of the spectrum under fast tumbling conditions. If there are no preferred molecular orientations, the appropriate average is (eqn 3.18)

$$\lambda R_{AX} \int_0^{\pi/2} (3\cos^2\theta - 1)\sin\theta \, d\theta, \tag{5.10}$$

where the volume element $\sin\theta$ takes into account the probability of finding an AX pair with an orientation between θ and $\theta + d\theta$. As noted in Section 3.8, this integral is exactly equal to zero. Sufficiently fast tumbling reduces the Pake pattern (Fig. 3.28) to a 'motionally narrowed' line whose width decreases as the tumbling rate increases.

We can recruit the theory of chemical exchange to estimate the linewidth and hence the T_2 of the motionally narrowed resonance. According to eqn 4.3, the linewidth in fast exchange is approximately $(\delta v)^2 \tau$ where δv is the difference in resonance frequencies of the exchanging sites and τ is the mean time spent in each site. By analogy, the dipolar linewidth for a molecule in solution should be roughly $R_{AX}^2 \tau_c$, i.e. the square of the dipolar coupling multiplied by the rotational correlation time of the motion. Thus, as the tumbling becomes faster, the dipolar interaction is more efficiently averaged, the line narrows, and T_2 increases.

For the random fields relaxation mechanism, theory (Slichter 1990) gives

$$\frac{1}{T_2} = \tfrac{1}{2}\gamma^2\langle B_{loc}^2\rangle J(\omega_0) + \tfrac{1}{2}\gamma^2\langle B_{loc}^2\rangle J(0). \tag{5.11}$$

$\lambda = 1$ for heteronuclear spins and $\lambda = \tfrac{3}{2}$ for homonuclear spins. See Section 3.8.

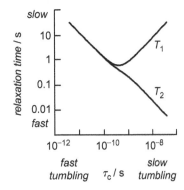

Fig. 5.11 The dependence of T_1 and T_2 on the rotational correlation time, τ_c, from eqns 5.4 and 5.11, when $\omega_0/2\pi = 400$ MHz. Both axes are logarithmic. The regions of the graph corresponding to fast and slow tumbling, and fast and slow relaxation are indicated. The value of $\gamma^2\langle B_{loc}^2\rangle$ used (4.5×10^9 rad^2 s^{-2}) is roughly appropriate for the dipolar coupling of two protons separated by 2 Å.

The first part of this expression is simply $\frac{1}{2}T_1^{-1}$ (eqn 5.4) and represents the life-time broadening caused by spin–lattice relaxation. The second term comes from the motional narrowing we have just been discussing, with $J(0) = 2\tau_c$ and $\gamma^2\langle B_{loc}^2\rangle \approx R_{AX}^2$.

The motional dependence of T_2 is shown in Fig. 5.11, together with the corresponding T_1 behaviour from eqn 5.4 and Fig. 5.5. As anticipated, T_2 increases as the tumbling gets faster and more effectively averages the dipolar coupling. In the slow motion regime ($\omega_0\tau_c \gg 1$) the component of T_2 arising from spin–lattice relaxation is negligible (because $J(\omega_0) \ll J(0)$, see Fig. 5.4) and the spin–spin relaxation rate is simply proportional to the correlation time (because $J(0) = 2\tau_c$).

The two relaxation times are identical in the extreme narrowing limit, where the tumbling is fast compared to the Larmor frequency ω_0: when $\omega_0\tau_c \ll 1$, $J(\omega_0)$ and $J(0)$ are equal and so, therefore, are the two terms in eqn 5.11. Comparing this with eqn 5.4, it is clear that $T_1 = T_2$.

Although almost everything said hitherto has been about the dipolar mechanism, there *are* other sources of relaxation. Any magnetic interaction experienced by nuclear spins can, in principle, lead to relaxation provided random molecular motions are able to produce an appropriate time dependence. Examples include modulation of anisotropic chemical shifts by molecular tumbling; modulation of J-couplings by internal motion; and interactions with the strong fields generated by electron spins in paramagnetic molecules. For $I > \frac{1}{2}$ nuclei, however, there is an additional, often dominant, relaxation mechanism to which we now turn.

5.7 Quadrupolar relaxation

A nucleus with a spin quantum number greater than $\frac{1}{2}$ possesses an *electric quadrupole moment* in addition to its magnetic dipole moment. One can think of this in terms of an ellipsoidal charge distribution in the nucleus (Fig. 5.12) with (for example) an excess of positive charge near the north and south poles and a corresponding depletion around the equator. Unlike electric dipoles—for example a polar molecule such as HCl—electric quadrupoles do not interact with spatially uniform electric fields, but only with *electric field gradients*, a property that may be understood by regarding the nuclear quadrupole moment as two identical back-to-back electric dipoles (Fig. 5.13).

The electric field gradient at a nucleus is a measure of the *non-uniform* distribution of local electronic charge: in sufficiently high symmetry environments—spherical, cubic, octahedral, or tetrahedral—the electric field gradients generated by surrounding charges exactly cancel out, giving no net quadrupolar interaction. Nuclei in lower symmetry environments, however, experience non-zero electric field gradients which depend on the orientation of the molecule in the magnetic field of the NMR spectrometer. For example, consider a molecule containing two regions of negative charge at a fixed distance from a quadrupolar nucleus (Fig. 5.12). These charges have a more favourable electrostatic interaction with the non-spherical nucleus when the molecule is oriented such that

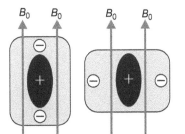

Fig. 5.12 A quadrupolar nucleus, whose non-uniform charge distribution is represented by a prolate ellipsoid, in the electric field of two negative charges in a molecule, shown as a box. As the molecule rotates, the spin axis remains aligned along the B_0 field, and the energy of the nucleus is modulated as indicated.

they are closer to its poles than to its equator. This interaction, like the magnetic dipolar coupling, produces splittings in the NMR spectra of single crystals, and gives broad lines for powders and disordered solids. Quadrupolar interactions also cause *spin relaxation* when modulated by molecular tumbling.

Many $I > \frac{1}{2}$ nuclei in low symmetry environments have large quadrupolar interactions and consequently very efficient spin relaxation. The most obvious manifestations of this are the large linewidths (small T_2) generally observed for nuclei such as ^{14}N, ^{17}O, ^{35}Cl, ^{37}Cl. For example, the ^{14}N resonance of the pyramidal NMe_3 is almost 100 Hz wide, while $^{14}NMe_4^+$, $^{15}NMe_3$, and $^{15}NMe_4^+$ all have nitrogen linewidths less than 1 Hz ($^{14}NMe_4^+$ is tetrahedral and has no electric field gradient at the position of the nitrogen, and ^{15}N has $I = \frac{1}{2}$ and therefore no quadrupole moment). Quadrupolar line broadening severely reduces resolution, and accounts for the relative unpopularity of $I > \frac{1}{2}$ nuclei in liquid-state NMR.

The other major consequence of quadrupolar relaxation is the loss of multiplet structure for spins that are J-coupled to quadrupolar nuclei. Efficient spin–lattice relaxation of, for example, an $I = 1$ nucleus causes it to flip rapidly between its three spin states ($m = +1, 0, -1$) so that a coupled spin (Larmor frequency v_0, coupling constant J) has a resonance frequency that jumps rapidly between $v_0 + J$, v_0, and $v_0 - J$. If this happens at a rate that is fast compared to the frequency differences involved, i.e. J and $2J$, then one should see just a *single line* at the mean frequency, v_0 instead of a 1:1:1 triplet (see the discussion of chemical exchange effects in Chapter 4). This is essentially identical to the collapse of multiplet splittings brought about by fast intermolecular proton exchange (Fig. 4.12). Examples of molecules in which quadrupolar nuclei do not lead to multiplet splittings can be seen in Fig. 3.13 (^{14}N, ^{79}Br, and ^{81}Br), and Fig. 4.10 (^{99}Ru and ^{101}Ru).

5.8 Examples–structure

The applications of relaxation effects in NMR usually fall into one of two categories: structure and dynamics. Roughly speaking, T_1 measurements are more useful for motional studies, and NOEs for structural investigations, although there is plenty of overlap. In general, to get information on the structure and/or conformation of a molecule, one needs to have some knowledge of, or make assumptions about, its dynamical properties, and vice versa. For example, it is difficult to use NOEs to determine the average conformations of highly flexible molecules.

Quaternary carbons

The ^{13}C spin–lattice relaxation times of organic molecules are often dominated by ^{13}C-^1H dipolar interactions, the closest protons having the greatest effect because of the r^{-6} distance dependence (eqn 5.5). Quaternary carbons, which have no directly bonded protons, therefore usually relax more slowly than CH and CH_2 carbons in the same molecule, at least in the absence of appreciable internal motion. For example, the four types of quaternary carbons in fluoranthene (Fig. 5.14) relax five to eight times slower than the five methine carbons.

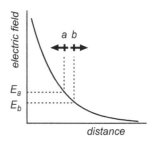

Fig. 5.13 The interaction of an electric quadrupole, regarded as two barely separated back-to-back dipoles, with a non-uniform electric field. The net interaction is proportional to the *difference* in electric fields experienced by the two dipoles ($E_a - E_b$) which, in turn, is determined by the *gradient* of the electric field at the position of the quadrupole.

	T_1	NOE	
a	2.4	1.73	
b	2.4	1.98	
c	2.1	1.86	methine
d	2.5	1.72	
e	2.2	1.86	
f	16.1	0.33	
g	15.6	0.34	quaternary
h	15.1	0.15	
i	11.6	0.33	

Fig. 5.14 ^{13}C spin–lattice relaxation times (in seconds) and $^{13}C\{^1H\}$ nuclear Overhauser enhancements for fluoranthene. (Data from C. Yu and G. C. Levy, *J. Am. Chem. Soc.,* **106** (1984) 6533.)

A similar picture emerges from the $^{13}C\{^1H\}$ NOEs in fluoranthene (Fig. 5.14). The enhancements for the methine carbons are all close to the maximum value of 1.99 as expected for a predominantly dipolar relaxation mechanism (eqn 5.8). The smaller NOEs, for the quaternary carbons, are consistent with slower dipolar relaxation and a relatively greater importance of other mechanisms which do not cause cross relaxation. Carbons *f*, *g*, and *i* are enhanced by protons on adjacent carbons (*d*, *a*, and *c*, respectively), while *h* must rely on even more distant protons, hence its small NOE.

CH and CH$_2$ carbons

In a rigid molecule, a methylene carbon with dipolar interactions to two directly bonded protons should relax *twice* as fast as a methine carbon bearing a single proton, other things being equal. For example, the ratio of ^{13}CH to $^{13}CH_2$ spin–lattice relaxation times in adamantane (Fig. 5.15) is indeed close to 2.0. The actual ratio of 1.8 may be understood by considering the different number of β protons (on adjacent carbons) in the two cases: six for the methines and two for the methylenes. The extra dipolar relaxation due to the β protons is thus three times more effective for the CH carbons, causing them to relax more rapidly than would be expected simply on the basis of the number of α protons. A simple calculation using standard bond lengths and angles shows that the T_1 ratio (CH:CH$_2$) should be reduced from 2.0 to 1.82 in excellent agreement with the experimental value. Finally it must be said that ^{13}C spin–lattice relaxation is not always as straightforward as it is in adamantane. In more complex molecules of lower symmetry, anisotropic tumbling and internal motions (see Section 5.9) often serve to complicate matters.

The r^{-6} distance dependence of the dipolar relaxation mechanism is an invaluable source of information on the structures and conformations of molecules. By far the most useful relaxation parameter is the NOE which under the right conditions is governed by a single relaxation mechanism (dipolar) and by a single internuclear distance. Spin–lattice relaxation times are less informative, being determined by the dipolar interactions to all nearby nuclei, as well as by other relaxation mechanisms.

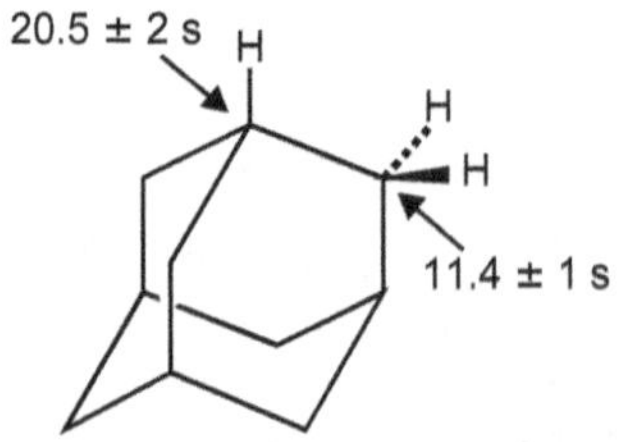

Structural information from NOEs

A difficulty routinely faced by chemists is to decide whether a newly synthesized molecule is the desired compound or a closely related one. In many cases, the structure will be known except for some small but crucial detail like the stereochemistry or the positioning of a substituent. In many cases, the NOE provides a quick and unambiguous answer.

Trimethylsilylation of the triisopropylsilylindole in Fig. 5.16 can lead to two possible trimethylsilyl derivatives, with either 4 or 7 substitution. Both isomers are expected to show four doublets and one doublet of doublets in the aromatic region of the ^1H spectrum. Unless one were happy to rely on subtle chemical shift arguments, it would be difficult to decide which compound had been formed.

20.5 ± 2 s

11.4 ± 1 s

Fig. 5.15 ^{13}C spin–lattice relaxation times (in seconds) of the CH and CH$_2$ carbons of adamantane. (Data from K. F. Kuhlmann, D. M. Grant, and R. K. Harris, *J. Chem. Phys.*, **52** (1970) 3439.)

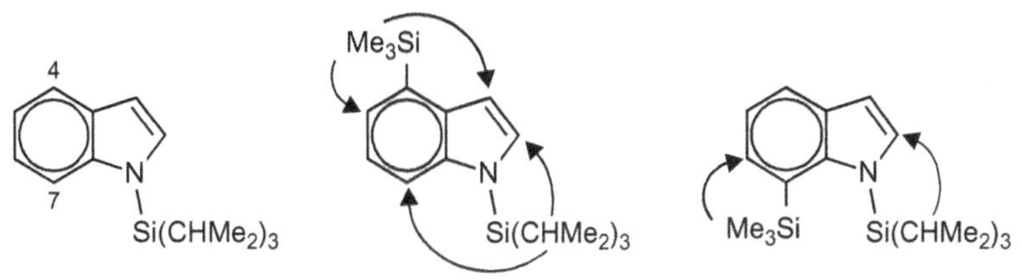

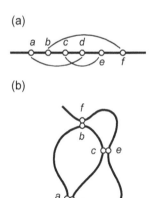

Fig. 5.16 Two possible products of trimethylsilylation of *N*-triisopropylsilylindole (left) showing the ^{1}H–^{1}H nuclear Overhauser effects that would result from saturation of the SiMe$_3$ methyl protons and the Si(CHMe$_2$)$_3$ methine proton. (G. Nechvatal and D. A. Widdowson, *J. Chem. Soc. Chem. Commun.*, (1982) 467.)

The solution is provided by two quick ^{1}H–^{1}H NOE experiments: irradiation of the trimethylsilyl singlet boosts the intensity of two of the four ^{1}H doublets, while irradiation of the septet from the CH proton in the triisopropylsilyl group enhances the *other* two doublets. These observations are only consistent with 4-substitution. The 7-trimethylsilyl derivative would have shown enhancements for just one doublet in each experiment.

Protein structure determination

One of the most impressive advances in NMR in recent years has been the determination of the three-dimensional structures of proteins in solution. Starting from the sequence of amino acids and a set of NMR spectra it is now possible to define the conformations of proteins containing up to 200 amino acid residues, with relative molecular masses up to about 25,000.

The trick is to detect many hundreds of NOEs between pairs of protons at known positions in the molecule. Some of these will link nuclei in the same residue, or in neighbouring residues in the sequence, but many will connect protons in very different parts of the molecule. These 'connectivities' give upper limits on the distances between protons in the range 2–6 Å depending on the size of the protein. Even though these distance constraints are short-range compared to the overall dimensions of the protein (typically a few tens of angstroms), and often rather imprecise, if there are enough of them distributed throughout the protein they define the structure very effectively (Fig. 5.17). All that is required is a sophisticated computer algorithm to search for the conformations that satisfy all the constraints, together with restrictions on torsion angles from three-bond *J*-couplings (Section 3.7) and residual dipolar couplings (Section 3.8).

The whole procedure relies on having first *assigned* a substantial fraction of the ^{1}H spectrum: i.e. as many lines as possible must be attributed to specific protons in the molecule. The initial stage of this formidable task is to identify NMR lines corresponding to protons in the *same* amino acid residue. This can be done by mapping out the network of *J*-couplings in each residue, knowing that protons only have appreciable coupling constants if they are separated by two or three chemical bonds. For example, alanines are easily identified because Ala is the

Fig. 5.17 Protein structure determination from nuclear Overhauser effects (schematic). NOEs are detected between nuclei at different positions along the polypeptide chain (a), which must therefore be close to one another. A computer search algorithm finds structures which satisfy all these distance constraints (b).

only amino acid whose αCH proton and side-chain (a methyl group) make up an AX_3 spin system. The next stage is to link together the individual amino acid spin systems by detecting NOEs between protons in consecutive residues, e.g. between the backbone amide NH protons, which are separated by 2–5 Å. In this way the spectrum can be assigned sequentially, stepping from one residue to the next.

This method of assigning the ^{1}H spectrum becomes challenging for larger proteins (more than about 100 amino acids) where the slower tumbling causes faster spin–spin relaxation and broader lines, making it difficult to resolve ^{1}H–^{1}H J-couplings. A solution is to take advantage of heteronuclear J-couplings by using ^{13}C and ^{15}N isotopically enriched proteins (see Sections 6.6 and 6.7, Campbell (2012), and Cavanagh et al. (2007)).

Two kinds of NMR experiment are therefore required, to determine the *through-bond* (J-couplings) and *through-space* connectivities (NOEs). Although this sort of information can be obtained for small molecules by irradiating each resonance in turn and looking for changes elsewhere in the spectrum, this is not feasible for proteins which typically have many overlapping NMR lines. In practice, protein structure determination has been made possible by the development of *multi-dimensional* NMR techniques, in which resonances are spread out into a second or third frequency dimension, to alleviate overcrowding and to allow all connectivities of a particular type to be obtained in a single experiment without the need for frequency-selective irradiation. The basis of these techniques is outlined in Sections 6.6 and 6.7.

5.9 Examples—dynamics

As mentioned in Section 5.8 above, information on molecular motions can usually only be extracted from relaxation data when the structure is well defined. Many relaxation studies therefore use 'probes' of known geometry, with a dominant (usually dipolar) relaxation mechanism, for example ^{13}C in CH groups, ^{15}N in NH groups, etc.

Methyl group internal rotation

In one of the examples above, we saw that CH_2 carbons often relax about twice as rapidly as CH carbons. Extrapolating, we might expect that a methyl carbon which is relaxed by dipolar couplings to three directly bonded protons would relax three times as fast as a ^{13}CH group in the same molecule. However this is rarely the case because of the rapid internal rotation of methyl groups. For example, the methyl carbons in mesitylene (Fig. 5.18(a)) undergo dipolar relaxation *more slowly* than the ring carbons. This is easily understood when one remembers that for extreme narrowing, faster motion means slower relaxation. The ring carbons which enjoy no internal motion must rely on the relatively slow overall tumbling of the molecule to modulate the ^{13}C–^{1}H dipolar interactions and cause relaxation. The methyl groups, however, rotate much faster than the molecule tumbles and give more effective averaging of the dipolar interactions and hence slower relaxation.

In *ortho*-xylene (Fig. 5.18(b)), however, the methyl and methine ^{13}C T_1s are very similar because steric interactions between the adjacent methyl groups slow down the internal motion and accelerate methyl carbon relaxation relative to the ring carbons. Analysis of the data for *ortho*-xylene shows that the internal rotation is roughly twelve times faster than the overall tumbling and has a barrier height of about 6 kJ mol^{-1}.

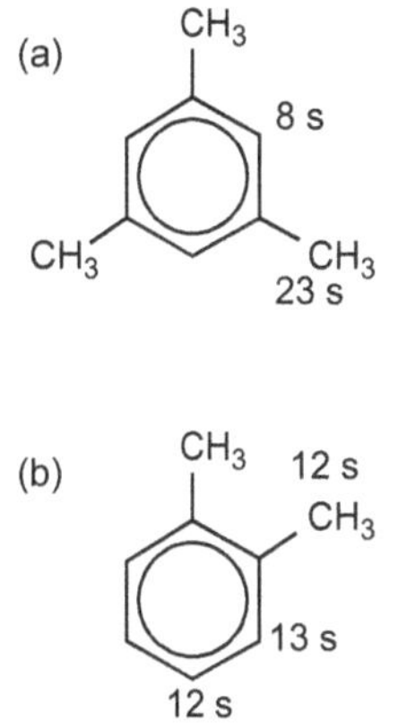

Anisotropic rotation

Figure 5.19 shows the ^{13}C spin–lattice relaxation times for the phenyl carbons in diphenyldiacetylene. From the foregoing discussion one might anticipate that the carbons in position 2 (C2) would relax a little more slowly than C3 and C4 because of the different numbers of nearby protons. In fact C2 and C3 have very similar T_1s, which are almost five times longer than the T_1 of C4.

The origin of this effect lies in the anisotropic tumbling of this rod-like molecule: the end-over-end motion is much slower than rotation around the long axis. The relaxation of C4 is predominantly caused by motions that modulate the dipolar coupling to its attached proton, i.e. by motions that change the angle between this C–H bond and the magnetic field direction. Clearly the rapid long-axis rotation is ineffective in this respect, so that it is the end-over-end motion that relaxes C4. Since this motion is slow, and the molecule falls in the extreme narrowing limit, the relaxation of C4 is fast. By contrast, the CH vectors of C2 and C3 point away from the long axis, allowing the rapid motion around this axis to modulate the C–H dipolar coupling, resulting in slow relaxation. The end-over-end motion is about twenty times slower than that around the long axis.

Of course this is a fairly extreme example of anisotropic motion: not all molecules deviate as much from spherical symmetry as this one. However, the effects of anisotropic motions can be discerned in many small molecules.

Alkyl chains

^{13}C spin–lattice relaxation can be used to study the motions of highly flexible molecules. For example, Fig. 5.20 shows T_1s for the carbons in the linear compounds decane ($C_{10}H_{22}$), eicosane ($C_{20}H_{42}$), and decan-1-ol ($C_{10}H_{21}OH$). These data can be understood in terms of the *overall motion* of the molecule, which becomes slower as the molecular mass and viscosity increase, and the *internal motion* due to rotation about individual C–C bonds.

Comparing the two alkanes, eicosane tumbles slowly, hence the generally short T_1s, and has relatively rapid internal motion which causes a strong variation in T_1 along the chain, with the slowest relaxation at the mobile chain ends. In decane, which tumbles more rapidly, the two types of motion occur at more similar rates, leading to a reduced importance of internal motion and a smaller spread in T_1s.

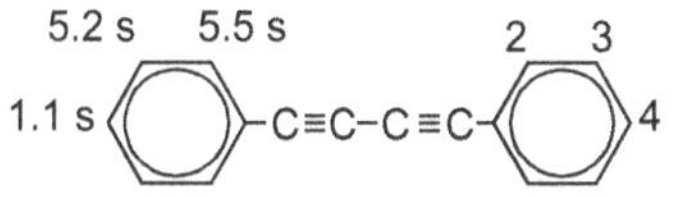

Fig. 5.18 ^{13}C spin-lattice relaxation times of CH and CH$_3$ carbons in (a) mesitylene and (b) *ortho*-xylene. Only the dipolar contributions to the relaxation times are shown. (Data from K. F. Kuhlmann and D. M. Grant, *J. Chem. Phys.*, **55** (1971) 2998.)

Fig. 5.19 ^{13}C spin-lattice relaxation times of the C2, C3, and C4 ring carbons in diphenyldiacetylene. (Data from G. C. Levy, J. D. Cargioli, and F. A. L. Anet, *J. Am. Chem. Soc.*, **95** (1973) 1527.)

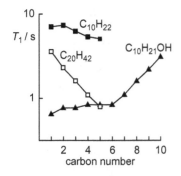

Fig. 5.20 ^{13}C spin-lattice relaxation times for the alkyl chains of decane, eicosane, and decan-1-ol. The vertical scale is logarithmic. (Data from D. Doddrell and A. Allerhand, *J. Am. Chem. Soc.*, **93** (1971) 1558 and J. R. Lyerla, H. M. McIntyre, and D. A. Torchia, *Macromolecules*, **7** (1974) 11.)

In decanol, the T_1s are all shorter than for decane because of the increased viscosity, and decrease steadily towards the OH group, reflecting the restriction of internal motion caused by hydrogen bonding.

Solid benzene

The ^{1}H NMR spectrum of solid benzene below 90 K consists of a single line some 40 kHz wide, the result of a multitude of strong unresolved intra- and intermolecular dipolar splittings. As shown in Fig. 5.21(a), the line sharpens dramatically between 90 and 120 K, and then remains more or less unchanged up to the melting point, where rapid tumbling reduces the width to less than 1 Hz (not shown). The linewidth change in the solid is due to molecular reorientation around the six-fold axis of the molecule which partially averages the dipolar interactions. Narrowing occurs when the frequency of rotation becomes comparable to the 40 kHz linewidth.

Further evidence of this motion comes from the proton T_1 of benzene (Fig. 5.21(b)) which goes through a minimum at 170 K. Evidently the average rotation frequency matches the ^{1}H resonance frequency (23.4 MHz in this case) at this temperature (see Fig. 5.5).

Use of selectively deuterated benzenes allows the intra- and intermolecular contributions to the relaxation to be separated and gives an intramolecular H–H distance of 2.495 ± 0.018 Å, in fine agreement with the structure of benzene determined by X-ray crystallography. The activation energy and pre-exponential factor for the spinning of the benzene molecules come out at 15.5 kJ mol^{-1} and 9.1×10^{12} s^{-1}.

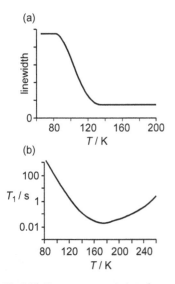

Fig. 5.21 Temperature variation of (a) the ^{1}H linewidth, and (b) the ^{1}H spin-lattice relaxation time of solid benzene. (After E. R. Andrew and R. G. Eades, *Proc. Roy. Soc. A*, **218** (1953) 537.)

5.10 Summary

- Nuclear spins attain thermal equilibrium by the process known as spin relaxation.

- Spin–lattice relaxation drives the populations of energy levels towards equilibrium.

- Spin–spin relaxation results in NMR line broadening.

- The spin–lattice and spin–spin relaxation times are known as T_1 and T_2.

- Spin relaxation is caused by the modulation of spin interactions by molecular motions.

- Dipolar coupling, rendered time-dependent by molecular tumbling, is a common source of spin relaxation.

- Dipolar coupling also causes cross relaxation which leads to the nuclear Overhauser effect.

- Quadrupolar nuclei often relax very rapidly.

- T_1, T_2, and the NOE give information on molecular structure and motion.

5.11 Exercises

1. The width of an NMR line is 0.1 Hz. What is its T_2?

2. How long (in multiples of T_1) is it necessary to wait for $\Delta n(t)$ in eqn 5.1 to relax to 99% of Δn_{eq}?

3. (a) Use eqns 5.3 and 5.4 to estimate the rotational correlation time, τ_c, that corresponds to the minimum ^{1}H T_1 on a 500 MHz spectrometer. (b) What is the minimum 500 MHz ^{1}H T_1 when $\gamma^2 \langle B_{loc}^2 \rangle = 4.5 \times 10^9$ rad^2 s^{-2}?

4. (a) The ^{13}C T_1 of an isolated CH group in a small rigid molecule ($\tau_c = 50$ ps) is 0.931 s. Use eqn 5.5 to estimate the C–H bond length. (b) Would you expect the ^{15}N T_1 of an isolated NH group in the same molecule to be longer or shorter than 0.931 s?

5. Which of the protons in 1,3-dinitrobenzene has (a) the shortest and (b) the longest spin–lattice relaxation time?

6. Which would you expect to have the larger ^{15}N linewidth: an amide nitrogen in a peptide ($\tau_c \approx 200$ ps) or in a protein ($\tau_c \approx 20$ ns)?

7. The ratio $T_1/T_2 = 10$ was measured for a ^{1}H resonance on a 600 MHz spectrometer. Use eqns 5.3, 5.4, and 5.11 to obtain a value for τ_c.

8. How would Fig. 5.5 differ for ^{1}H nuclei in an 800 MHz spectrometer?

9. Verify that $W_{\beta\alpha}/W_{\alpha\beta}$ must equal $\exp(\Delta E/k_B T)$ in order for spin–lattice relaxation to return the spins to thermal equilibrium. W_{jk} is the rate constant for relaxation from state j to state k.

10. The ^{1}H–^{1}H NOE vanishes when the two cross relaxation rates, W_2^{AX} and W_0^{AX}, are equal. Use $W_2^{AX}/W_0^{AX} = 6J(2\omega_0)/J(0)$ to find the value of τ_c at which this occurs on a 600 MHz spectrometer.

Answers to the exercises are provided at the back of the book. Full worked solutions are available on the Online Resource Centre at www.oxfordtextbooks.co.uk/orc/hore_nmr2e/

6 NMR experiments

6.1 Introduction

This final chapter attempts to explain how NMR experiments work. Modern NMR spectroscopy is much more than simply recording a spectrum and interpreting the positions, widths, and intensities of the lines. The spins can be manipulated to tailor the information that appears in the spectrum. For example, there are experiments designed to identify pairs of nuclei separated from one another by a small number of chemical bonds or by a few angstroms in space, or which enhance the NMR intensities of insensitive, low-γ nuclei by transferring magnetization to them from high-γ spins. For those who want to discover more than is presented in this chapter about the inner workings of such experiments and to learn about the quantum mechanics required to understand them properly, we recommend Hore, Jones, and Wimperis (2015).

We start by introducing the 'vector model'. Although it has its origin in the quantum mechanics of spin angular momentum, it has the distinct advantage of being pictorial and essentially non-mathematical. The disadvantage is that it only helps one understand the simplest NMR experiments. In Sections 6.4 and 6.5, we use the vector model to discuss two techniques for measuring spin relaxation times and an important method for sensitivity enhancement. First, a brief summary of the principles of practical NMR.

6.2 Experimental principles

In this section, we give an extremely superficial description of an NMR spectrometer. The intention is to provide just enough information to make the rest of the chapter intelligible. There are excellent textbooks that give far more insight into the practice of NMR spectroscopy (e.g. Freeman (1997a, 1997b, 2003), Berger and Braun (2004), Levitt (2008), Claridge (2009), and Keeler (2010)).

We consider the simplest NMR experiment, shown in Fig. 6.1, which starts with a short burst of radiofrequency (rf) radiation (known as a *pulse*) applied to the sample sitting in a strong magnetic field, B_0. The pulse excites the spins and produces an oscillating signal—the *free induction decay*—from which the spectrum is obtained.

As we saw in Chapter 1, the $\mathbf{B}_0$ field is responsible for producing the energy-level splitting and population differences necessary to do spectroscopy. Typical field strengths are in the range 2.35–23.5 T, giving ^{1}H Larmor frequencies $|\nu_0|$ =100–1000 MHz. To be suitable for NMR, the static magnetic field must be strong, homogeneous, and stable. Strong fields have three principal advantages: (a) higher sensitivity—the strength of the NMR signal increases with the energy-level spacing, eqn 1.12, and hence with B_0; (b) reduced overlap of multiplets—the frequency differences arising from chemical shifts increase linearly with B_0, while J-couplings are independent of B_0; (c) less severe strong coupling effects (Section 3.5). An exceedingly *uniform* field is required for the best possible spectral resolution. Spatial variations in the field experienced by the sample lead to a distribution of resonance frequencies and unwelcome line broadening (Fig. 6.2). Field homogeneity better than one part in 10^9 is required to get 1 Hz linewidths on a high-field spectrometer (e.g. 600 MHz). Finally, the field must not drift by more than one part in 10^9 during the course of an NMR experiment, which may take anything from a few seconds to several days.

These stringent specifications are best met by superconducting solenoids—coils of resistance-free alloy carrying a persistent current. At the time of writing, the strongest commercially available superconducting NMR magnet has a field of 23.5 T, corresponding to a ^{1}H Larmor frequency of 1000 MHz.

We turn now to the radiofrequency pulse (Fig. 6.1) and consider an NMR experiment on identical, non-interacting spin-$\frac{1}{2}$ nuclei with a Larmor frequency $\omega_0 = 2\pi\nu_0 = -\gamma B_0(1-\sigma)$ (eqn 2.5). The sample, typically ~1 ml of liquid in a ~5 mm diameter cylindrical glass tube, sits in the middle of the magnet, surrounded by the transmitter/receiver coil. The spectrometer sends a voltage pulse to the coil at a frequency (ω_{rf}) close to ω_0 which generates an oscillating magnetic field $\mathbf{B}_1(t)$ at the position of the sample. This radiofrequency field is perpendicular to and much weaker than $\mathbf{B}_0$ (typical strengths for ^{1}H NMR are ~1 mT) and is normally present for ~10 µs. More complex NMR experiments require *pulse sequences* comprised of several radiofrequency pulses whose durations, strengths, phases, frequencies, and timings are precisely controlled by the spectrometer.

The radiofrequency pulse leads to a transient nuclear magnetization that oscillates at the Larmor frequency of the spins and typically dies away within about a second. It produces an oscillating voltage in the coil which is detected by the spectrometer where it is 'mixed' with a reference signal, often at the same frequency as the pulse. The resulting signal is a damped oscillation at frequency $\Omega = \omega_0 - \omega_{rf}$, known as the *free induction decay* (Fig. 6.1), which is processed in the spectrometer to obtain the spectrum. Thus, NMR data are acquired in the *time-domain* and transformed into the *frequency domain* to obtain a spectrum.

6.3 Vector model

The preceding section raises several fundamental questions about the theory and practice of NMR. How does a monochromatic (single frequency) pulse excite spins without necessarily being exactly on resonance (i.e. $\omega_{rf} = \omega_0$)? How

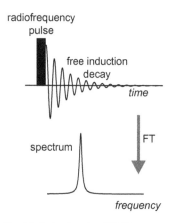

Fig. 6.1 A simple pulsed NMR experiment in which the spectrum is obtained from the free induction decay by Fourier transformation (FT).

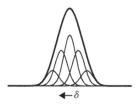

Fig. 6.2 The broadening of an NMR line resulting from spatial inhomogeneity of the static magnetic field B_0. Signals are drawn for *five* regions of the sample experiencing different values of B_0. The amplitude of each contribution is proportional to the number of spins in that region. The bold line represents the lineshape observed for the sample as a whole. In reality there is a continuous, rather than discrete, distribution of B_0 values and therefore resonance frequencies.

does this lead to a nuclear magnetization that oscillates at the Larmor frequency, ω_0? How is the spectrum obtained from the free induction decay? What is the advantage of using pulsed (rather than continuous) radiofrequency fields? The answers, together with insight into the operation of more complex NMR experiments, can be had from the *vector model*.

Rotating frame

The total magnetization of a large number of identical, non-interacting spin-$\frac{1}{2}$ nuclei (magnetogyric ratio γ) can be represented by a vector $\boldsymbol{M}$. Although $\boldsymbol{M}$ arises from the magnetic moments of the individual spins which obey *quantum mechanics*, its motion can be described using *classical mechanics*.

In a static magnetic field $\boldsymbol{B}$ the magnetization vector $\boldsymbol{M}$ *precesses* around the field direction in the same way the axis of a gyroscope or a spinning top precesses around the Earth's gravitational field (Fig. 6.3). The angular frequency of this motion is $-\gamma B$, i.e. $\gamma B/2\pi$ cycles per second. The stronger the magnetic field and the larger the magnetogyric ratio, the faster the precession.

Two distinct magnetic fields determine the motion of $\boldsymbol{M}$ in an NMR experiment: the strong static field $\boldsymbol{B}_0$, and the much weaker radiofrequency field, $\boldsymbol{B}_1(t)$. The static field vector $\boldsymbol{B}_0$ is taken to be parallel to the z'-axis in an (x',y',z') coordinate system conventionally called the *laboratory frame* (otherwise known as the real world). The radiofrequency field generated by the alternating current in the coil can be considered to rotate around the z'-axis in the $x'y'$-plane at frequency ω_{rf}. The total, or *effective*, field $\boldsymbol{B}_{eff}(t)$ present during a radiofrequency pulse (the vector sum of $\boldsymbol{B}_0$ and $\boldsymbol{B}_1(t)$) is thus tilted slightly away from the z'-axis, and precesses around it at frequency ω_{rf} (Fig. 6.4(a)).

When the radiofrequency field is not present, the effective field $\boldsymbol{B}_{eff}$ is simply $\boldsymbol{B}_0$. The magnetization vector $\boldsymbol{M}$ therefore precesses around the z'-axis at the Larmor frequency $\omega_0 = -\gamma B_0$, or $\omega_0 = -\gamma B_0(1-\sigma)$ when the chemical shift is included. The situation is more complicated when the radiofrequency field is

We assume $\gamma > 0$ here to avoid having to use $|\gamma|$ everywhere.

In fact the radiofrequency field generated by the alternating current in the coil oscillates along (say) the x'-axis in the laboratory frame with peak amplitude $2B_1$. Defining $\boldsymbol{x'}$ and $\boldsymbol{y'}$ as unit vectors along the x' and y' axes, this linearly polarized field can be written $2B_1\cos\omega_{rf}t\,\boldsymbol{x'} = B_1(\cos\omega_{rf}t\,\boldsymbol{x'} + \sin\omega_{rf}t\,\boldsymbol{y'}) + B_1(\cos\omega_{rf}t\,\boldsymbol{x'} - \sin\omega_{rf}t\,\boldsymbol{y'})$, that is as the sum of two fields of amplitude B_1 rotating in opposite senses (one clockwise, the other anticlockwise) around the z'-axis in the $x'y'$-plane. Only the component that rotates in the same sense as the Larmor precession has a significant effect on the spins. The other can be safely ignored.

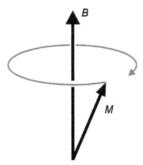

Fig. 6.3 The motion of the magnetization vector $\boldsymbol{M}$ in a magnetic field $\boldsymbol{B}$. $\boldsymbol{M}$ precesses around $\boldsymbol{B}$ at an angular frequency $-\gamma B$. The length of $\boldsymbol{M}$ and its projection onto $\boldsymbol{B}$ remain unchanged in the absence of spin relaxation. The diagram has been drawn for nuclei that have $\gamma > 0$ and hence a negative precession frequency.

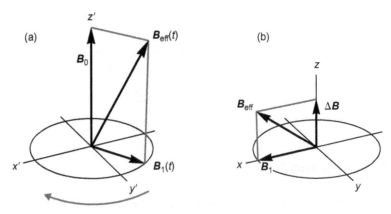

Fig. 6.4 The magnetic fields present in an NMR experiment in (a) the laboratory frame and (b) the rotating frame.

present because M precesses around B_{eff} which is itself precessing around the z'-axis. However, we can simplify things by imagining ourselves to be rotating around the z'-axis in step with $B_1(t)$, i.e. at frequency ω_{rf}. In this *rotating frame*, with coordinates (x,y,z), B_1 is stationary and has a direction in the xy-plane determined by the phase of the radiofrequency field. The other consequence of the transformation into the rotating frame is that the strength of the magnetic field along the z-axis changes from B_0 to

$$\Delta B = B_0 + \omega_{rf}/\gamma = -(\omega_0 - \omega_{rf})/\gamma. \tag{6.1}$$

This comes about because the precession frequency of M in the rotating frame is

$$\Omega = \omega_0 - \omega_{rf}. \tag{6.2}$$

For example, when the radiofrequency field is on-resonance ($\omega_{rf} = \omega_0$), $\Omega = 0$ and M is stationary in the rotating frame (when $B_1 = 0$). Ω and ΔB are known as the offset frequency and offset field, respectively, and are related by $\Omega = -\gamma \Delta B$. 'Offset' here refers to the difference between the Larmor frequency and the radiofrequency.

To sum up, M precesses in the rotating frame around the vector sum of two static fields: B_1 in the xy-plane and ΔB along the z-axis (Fig. 6.4(b)). Although B_0 is always much stronger than B_1, the offset field ΔB can be smaller than B_1 if ω_{rf} and ω_0 are not too different and B_1 is large enough. As we shall see, NMR experiments are usually arranged so that $B_1 \gg \Delta B$.

The rotating frame concept is exceedingly convenient not only because it removes the unpleasant need to think about time-dependent fields, but also because NMR spectrometers actually detect offset frequencies (eqn 6.2) as mentioned at the end of Section 6.2. From now on, our discussion will be based exclusively in the rotating frame.

Thermal equilibrium

We are now ready to consider what happens during the experiment shown in Fig. 6.1. Prior to the pulse, everything is at thermal equilibrium. The nuclear spins ($\gamma > 0$) are polarized by the B_0 field such that the projections of their magnetic moments onto the z-axis are slightly more likely to be positive (lower energy) than negative (higher energy). The components of the individual magnetic moments onto the x' and y'-axes, however, are completely random because B_0 has no x' or y' component that could produce a transverse polarization. The magnetization vector M therefore points initially along the positive z-axis. The total magnetization at equilibrium is $M = \frac{1}{2}\gamma\hbar\Delta n_{eq}$, where Δn_{eq} is the population difference given by the Boltzmann distribution. More generally, $M_z(t)$, the projection of M onto the z-axis at any time t during an experiment, is equal to $\frac{1}{2}\gamma\hbar\Delta n(t)$ where $\Delta n(t)$ is the population difference at that time.

Radiofrequency pulses

When the radiofrequency field is switched on, M starts to precess around B_{eff} in the rotating frame. As mentioned above, ω_{rf} and B_1 are normally chosen so that

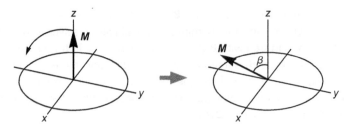

Fig. 6.5 A radiofrequency pulse causes the magnetization vector **M** to precess in the rotating frame around an axis in the *xy*-plane. The rotation angle β is given by eqn 6.3. As drawn, the phase of the pulse is +*x* so that **M** is rotated towards the −*y*-axis.

$B_1 \gg \Delta B$ which means that the effective field is simply B_1. The pulse causes **M** to precess around B_1 at frequency $\omega_1 = -\gamma B_1$ (Fig. 6.5). The angle through which the magnetization turns is called the flip angle, β:

$$\beta = -\omega_1 t_p = \gamma B_1 t_p \tag{6.3}$$

in which t_p is the duration of the pulse in seconds and β is in radians. The most commonly used pulses have flip angles, $\beta = \frac{1}{2}\pi\,(90°)$ or $\beta = \pi\,(180°)$. A 90° pulse rotates the equilibrium magnetization **M** from the *z*-axis into the *xy*-plane and gives the maximum possible transverse magnetization. When the phase of the pulse is *x*, this rotation leaves **M** pointing along the −*y*-axis (Fig. 6.5). When the phase of the pulse is *y*, **M** ends up on the +*x*-axis. A 180° pulse rotates **M** from the +*z*-axis to the −*z*-axis whatever the phase of the pulse. Different flip angles are achieved by appropriate choices of B_1 and t_p (eqn 6.3).

We can now see how radiofrequency pulses are able, *uniformly* and *simultaneously*, to excite nuclei that have different chemical shifts, i.e. different offset frequencies, Ω. Consider a 400 MHz ^{1}H experiment on a sample containing protons with chemical shifts in the range $0 < \delta < 10$ ppm. If ω_{rf} is chosen to match the Larmor frequency of spins in the middle of the spectrum ($\delta \approx 5$ ppm) then the offset frequencies will lie in the range $-2\ \text{kHz} < \Omega/2\pi < +2\ \text{kHz}$ (i.e. ±5 ppm × 400 MHz) and the offset fields will be $|\Delta B| < 47\ \mu\text{T}$ (using $\Omega = -\gamma \Delta B$). Now suppose we excite the spins with a 90° pulse with $t_p = 5\ \mu\text{s}$. The strength of this radiofrequency field is $B_1 \approx 1170\ \mu\text{T}$ (using eqn 6.3). Thus $B_1 \gg |\Delta B|$ is satisfied and the pulse will therefore rotate the magnetization vectors of *all* the protons in the sample onto the −*y*-axis, irrespective of their chemical shifts.

This example illustrates another feature of pulsed NMR experiments: although a radiofrequency pulse can excite all the protons in a sample it will not significantly affect other nuclides. For instance, the offset frequency ($\Omega/2\pi$) of the ^{13}C nuclei in the above example is roughly 300 MHz (the difference in the Larmor frequencies of ^{1}H and ^{13}C in a 9.4 T magnetic field). This is clearly so much larger than the value of $\gamma B_1/2\pi$ (50 kHz for a 5 μs 90° pulse) that the ^{1}H pulse will have a negligible effect on the carbons. To obtain a ^{13}C spectrum one would need to set ω_{rf} close to the ^{13}C Larmor frequency.

The nature of the excitation in an NMR experiment is a little unusual. A 180° pulse rotates **M** from the +*z*-axis to the −*z*-axis: its net effect is to invert the population difference, Δn_{eq}, by swapping the populations of the $m = \pm\frac{1}{2}$

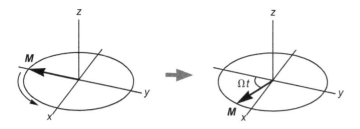

Fig. 6.6 Following a 90°_x pulse, the magnetization vector **M** precesses around the z-axis in the rotating frame at frequency $\Omega = -\gamma\Delta B = \omega_0 - \omega_{rf}$ (ignoring relaxation). **M** precesses through an angle Ωt radians in time t.

states. In contrast, a 90° pulse equalizes the two populations ($\Delta n_{eq} \to 0$) by converting the whole of the equilibrium z-magnetization into transverse magnetization in the xy-plane. This means that as soon as **M** is tipped away from the z-axis, the individual nuclear magnetic moments no longer have random phases in the xy-plane. This is known as *coherence*. This seemingly strange behaviour is a consequence of the quantum mechanical properties of the spins and is fundamental to the success of NMR (Hore, Jones, and Wimperis (2015)).

Free precession and the free induction decay

The next step is to discover what happens *after* a 90° pulse. As soon as $\mathbf{B}_1$ is switched off, the only field remaining in the rotating frame is the offset field $\Delta\mathbf{B}$, pointing along the z-axis. **M** therefore precesses in the xy-plane (Fig. 6.6) at frequency $\Omega = -\gamma\Delta B = \omega_0 - \omega_{rf}$. When the sample contains nuclei with different chemical shifts, each chemical environment gives rise to a magnetization vector that precesses in the rotating frame at the offset frequency appropriate to its chemical shift.

Up to now, we have ignored spin relaxation (Chapter 5). While this is normally an excellent approximation during the brief radiofrequency pulse, we cannot pretend that the spins will not return to equilibrium once the pulse has been switched off. There are two distinct relaxation processes: first, the recovery of the z-magnetization to its equilibrium value, which involves the re-establishment of the Boltzmann population difference, Δn_{eq} (e.g. eqn 5.1); and second, the decay to zero of the xy-magnetization, by randomization of the x and y components of the magnetic moments of the individual spins. The former is simply *spin–lattice relaxation*, which occurs with an exponential time constant T_1 (Section 5.2). More subtly, the latter is *spin–spin relaxation* (time constant T_2). In fact, the loss of phase coherence amongst the spins and the line broadening discussed in Section 5.6 are just different aspects of the same phenomenon. An analogy may help to clarify this point. A well-tuned bell, when struck, produces a single pure note which takes several seconds to die away. A poorly tuned bell, however, delivers a short lived 'clank' comprising a distribution of frequencies. Thus, prolonged ringing goes hand in hand with a narrow band of frequencies,

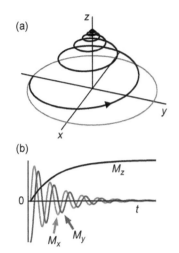

(a)

(b)

Fig. 6.7 (a) Following a 90°_x pulse, the magnetization vector **M** precesses around the z-axis and at the same time returns to its equilibrium position along the +z-axis. (b) The transverse components of **M** decay towards zero with characteristic time T_2, the spin–spin relaxation time. The z-component grows back to its equilibrium value with time constant T_1, the spin–lattice relaxation time. Both diagrams were drawn using $T_1 = T_2$ and are appropriate for identical, non-interacting spins.

The argument leading to Fig. 6.8 has been slightly simplified. In fact, the Fourier transform of $M_y(t) = A\cos \Omega_0 t$ using eqn 6.4 gives a line in the spectrum at $\Omega = -\Omega_0$ as well as one at $\Omega = \Omega_0$ because $\cos(-\Omega_0 t) = \cos(\Omega_0 t)$. In order to determine the sign of Ω_0 it is necessary to record both the x- and the y-component of the free induction decay. Treating them as the real and imaginary components of a complex signal, $M_{xy}(t) = M_y(t) - iM_x(t) = A\cos(\Omega_0 t) - iA\sin(\Omega_0 t) = A\exp(-i\Omega_0 t)$, and replacing $\cos(\Omega t)$ by $\exp(i\Omega t)$ in eqn 6.4, solves this problem. See Hore, Jones, and Wimperis (2015) for further information.

while rapidly damped oscillations correspond to a broad frequency response. So it is in pulse NMR experiments, where slow spin–spin relaxation (i.e. inefficient loss of coherence) leads to narrow NMR lines. The random local magnetic fields discussed in Chapter 5 produce small time-dependent variations in the precession frequencies of individual spins which lead to loss of phase-coherence in the sample as a whole.

Figure 6.7 shows the time dependence of **M** following a 90°_x pulse. The oscillating, decaying xy-magnetization is detected by the spectrometer via the voltage it induces in the receiver coil. More generally, this signal, known as the *free induction decay*, is the sum of the individual oscillating voltages arising from the different nuclear environments of the molecules in the sample, each with characteristic offset frequency (i.e. chemical shift and spin–spin couplings), amplitude, and T_2. It contains all the information necessary to obtain an NMR spectrum.

Fourier transformation

The final step is to unravel all the oscillating components in the free induction decay to obtain an NMR spectrum. This is done by means of a *Fourier transform*. In its simplest form the Fourier transform of the free induction decay, assumed to be proportional to the y-component of the magnetization, is:

$$S(\Omega) = \int_0^\infty M_y(t) \cos \Omega t \, dt, \tag{6.4}$$

where $S(\Omega)$ is the NMR intensity at offset frequency Ω. It is easy to see how this works (Fig. 6.8). Suppose the sample consists of identical isolated spins with offset frequency Ω_0 so that $M_y(t) = A\cos \Omega_0 t$, where the amplitude A is proportional to the number of spins. We ignore spin relaxation for the moment. When we multiply this signal by $\cos \Omega t$, as in eqn 6.4, with $\Omega \neq \Omega_0$, we get $A\cos \Omega_0 t \cos \Omega t$ which is an oscillating function with positive and negative values, whose integral over the range $0 \leq t \leq \infty$ is zero. Thus $S(\Omega \neq \Omega_0) = 0$. However, when $\Omega = \Omega_0$ the function to be integrated is $A\cos^2 \Omega_0 t$ which is always positive, so that $S(\Omega_0)$ is non-zero and proportional to A. Therefore Fourier transformation of a free induction decay comprising a single component $A\cos \Omega_0 t$ gives a spectrum $S(\Omega)$ that is everywhere zero except for a 'stick' of height A at frequency Ω_0.

As the Fourier transform is a linear operation, i.e.

$$FT[f(t) + g(t)] = FT[f(t)] + FT[g(t)], \tag{6.5}$$

the whole process works however many oscillatory components are present. For each contribution $A\cos \Omega_0 t$ in the free induction decay, the spectrum contains a stick of height A at frequency Ω_0. Finally, the effect of spin–spin relaxation is to cause the transverse magnetization to be exponentially damped which broadens the sticks in the spectrum. Figure 6.9 shows two examples of free induction decays and their Fourier transforms.

In practice, the spectrum is not obtained algebraically, as implied by eqn 6.4 and Fig. 6.8. Rather, the voltage that constitutes the free induction signal is first converted to digital form by sampling it at regular intervals and then subjected to a numerical Fourier transform using an efficient computer algorithm.

At this point it is worth asking why NMR signals are recorded using radio-frequency *pulses*. The early years of NMR spectroscopy (1945–1970) were dominated by *continuous wave* rather than *pulse* methods: spectra were obtained either by slowly varying B_0 in the presence of a fixed-frequency radiofrequency field, or by varying ω_{rf} keeping B_0 fixed, so as to bring spins with different chemical shifts sequentially into resonance, recording the NMR signal all the while. It was essential that the time taken to sweep through each NMR line was *long* compared to its T_1 and T_2, to avoid distorting the line shapes. This approach has now been superseded by the pulse techniques introduced above.

The principal advantage of pulsed NMR is *sensitivity*. Exciting and detecting the whole spectrum 'in one shot' is much more efficient than sweeping slowly through the spectrum, detecting one resonance at a time. For example, a continuous wave spectrum might take 100–1000 times longer to record than a single free induction decay. The time saved can be used to improve sensitivity, by recording and adding together several hundred or thousand free induction decays. The NMR responses, which are identical in every free induction decay, build up linearly with the number of signals recorded (N), while the inevitable instrumental noise varies randomly from one measurement to the next and adds up more slowly, as $\sqrt{N}$. The result of adding together N free induction decays is therefore an improvement in the signal-to-noise ratio of $N/\sqrt{N} = \sqrt{N}$ (Fig. 6.10). When N is large the improvement in sensitivity can be considerable. Mainly as a result of pulsed methods, NMR of ^{13}C at natural isotopic abundance is now routine and the detection of many low-sensitivity nuclei has become feasible.

The time-saving aspect of pulse NMR, known as the *multiplex* or *Fellgett advantage*, may be illustrated by considering how to determine the natural frequency of a bell. One approach would be to set up a frequency synthesizer and loudspeaker, and slowly tune the frequency through the audio range until the bell started to resonate, at which point the frequency could simply be read off from the synthesizer. This would be a time-consuming and tedious business. An obvious and more efficient method would simply be to strike the bell, and determine the frequency with which it oscillates using a microphone and oscilloscope or, if you have perfect pitch, just by listening. Evidently campanologists realized the merits of pulse excitation over continuous wave methods some time before NMR spectroscopists.

The other major advantage of pulse NMR is that it opens up the possibility of a vast number of new experiments in which the spectroscopist controls the information contained in the free induction decay by using *sequences* of pulses designed to excite the spins in a specific manner. To give just a glimpse of the huge number of variations on this theme, we start by looking at two simple experiments for determining the relaxation times T_1 and T_2, before moving on to something a bit more sophisticated.

(a)

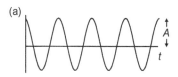

(b)

(c)

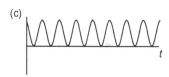

(d)

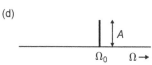

Fig. 6.8 A pictorial representation of the Fourier transform operation (eqn 6.4). (a) A model free induction decay oscillating at frequency Ω_0, as a function of time t (ignoring relaxation). (b) The product of (a) with $\cos\Omega t$ ($\Omega \neq \Omega_0$), which integrates to zero over the time interval $0 \leq t \leq \infty$. (c) The product of (a) with $\cos\Omega t$ ($\Omega = \Omega_0$), the integral of which is proportional to A, the amplitude of the free induction decay. (d) The NMR spectrum corresponding to (a).

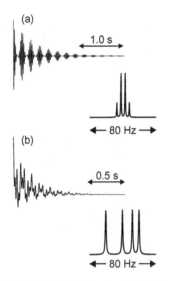

Fig. 6.9 Free induction decays and the corresponding spectra. (a) A quartet with a chemical shift offset of 40 Hz and $J=5$ Hz. $T_2=0.5$ s. (b) Four singlets with offset frequencies 0, 8, 20, and 40 Hz. $T_2=0.25$ s.

6.4 Relaxation time measurements

Inversion recovery—measurement of T_1

Spin–lattice relaxation times may be measured using the *pulse sequence* $180^\circ_x - \tau - 90^\circ_x$ (Fig. 6.11) comprising two pulses separated by an interval τ. The equilibrium magnetization (a) is inverted by the 180° pulse, leaving **M** along the $-z$-axis (b). During the delay τ, **M** undergoes partial spin–lattice relaxation (c) to give a z-magnetization, $M_z(\tau)$, which the 90° pulse rotates onto the y-axis (d). The free induction decay is recorded and Fourier transformed to give a spectrum containing peaks whose NMR intensities $S(\tau)$ are proportional to $M_z(\tau)$. The whole process is repeated for different values of τ so as to map out the recovery of the inverted magnetization (f). Assuming exponential relaxation,

$$S(\tau) = \left[1 - 2\exp(-\tau/T_1)\right]S(\infty). \tag{6.6}$$

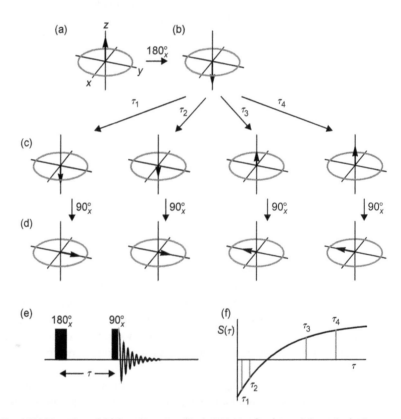

Fig. 6.11 Operation of the inversion recovery experiment for determining spin–lattice relaxation times. (a) Equilibrium. (b) After the 180°_x pulse. (c) After four different delays $\tau_1 < \tau_2 < \tau_3 < \tau_4$. (d) After the 90°_x pulse. (e) The pulse sequence. The subscripts indicate the phases of the two pulses. (f) The observed NMR signal intensity $S(\tau)$ as a function of the delay τ.

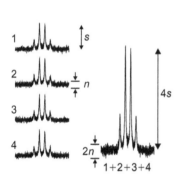

Fig. 6.10 The signal-to-noise ratio of an NMR spectrum can be improved by signal averaging. Co-addition of four independently recorded spectra, each with signal height $= s$ and noise standard deviation $= n$ gives a spectrum with signal $= 4s$ and noise $= 2n$. The signal-to-noise ratio is thus increased by a factor of two. More generally, the improvement is $\sqrt{N}$, where N is the number of signals added together.

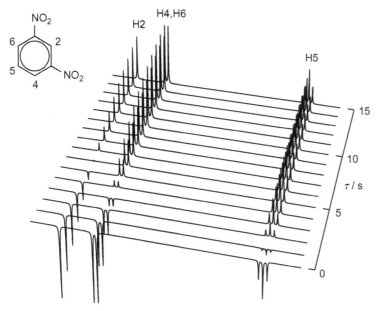

Fig. 6.12 ^{1}H T_1 measurement for 1,3-dinitrobenzene using the inversion recovery pulse sequence of Fig. 6.11. The T_1 values are 8.6 s (H2), 5.1 s (H4, H6), and 1.7 s (H5). The principal source of relaxation is the ^{1}H–^{1}H dipolar interactions between adjacent protons on the ring. The relaxation rates are in the order H5 > H4, H6 > H2 because the numbers of nearest neighbour protons are, respectively, 2, 1, and 0.

The T_1 of each peak in the spectrum can be obtained by plotting $\ln[S(\infty) - S(\tau)]$ against τ, where $S(\infty)$ is the fully relaxed NMR intensity obtained when $\tau \gg T_1$.

An example of an inversion recovery experiment is shown in Fig. 6.12.

Spin echoes—measurement of T_2

As discussed above, NMR line broadening and the damping of free induction decays arise from interactions that cause the spins to precess at slightly different frequencies, so destroying their phase-coherence. Two distinct processes are responsible: (a) the spin–spin relaxation induced by intra- or intermolecular magnetic fields and (b) the spatial inhomogeneity of the static field B_0 (Fig. 6.2). The latter is of no chemical interest but is an inevitable and unfortunate feature of most NMR experiments. As the width of an NMR line depends on both factors, spin–spin relaxation times cannot simply be obtained from linewidths (eqn 5.9). A technique that allows one to separate the two is the *spin echo* experiment, $90^\circ_x - \tau - 180^\circ_y - \tau -$, shown in Fig. 6.13. As in Fig. 6.11, the subscripts indicate the phases of the two pulses.

To understand the role played by field inhomogeneity, imagine the sample to be composed of a number of small regions, each of which experiences a different uniform field; spin–spin relaxation is ignored for the time being. The spins are assumed to be identical and non-interacting. After the 90° pulse (b), the total magnetization of each region precesses at a slightly different frequency

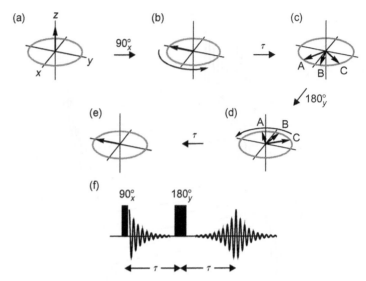

Fig. 6.13 Operation of the spin echo experiment for measurement of spin–spin relaxation times. (a) Equilibrium. (b) After the 90°_x pulse. (c) After the first τ delay. (d) After the 180°_y pulse. (e) After the second τ delay. (f) The pulse sequence. Relaxation is not included. The three magnetization vectors, labelled A, B, and C, arise from spins with different offset frequencies as a result of the spatial inhomogeneity of the B_0 field.

determined by the local magnetic field strength, so that the magnetization vectors fan out in the xy-plane Fig. 6.13(c). This dephasing attenuates the transverse magnetization of the sample as a whole. After a period τ, the 180°_y pulse flips the magnetization of each region around the y-axis to symmetrical positions in the xy-plane (d), whence precession continues for a further time τ. Whatever the precession frequencies, and whatever the value of τ, all regions come back into phase perfectly at the end of this second delay (e). This is called a *spin echo*: the dephasing caused by field inhomogeneity is said to be 'refocussed' by the 180° pulse and the NMR signal, which decayed rapidly after the 90° pulse, reappears at 2τ (f). The second half of the echo, which has the same form as the free induction decay, is detected and Fourier transformed to give a spectrum containing NMR lines whose *intensities* are independent of the field inhomogeneity.

Now consider the effect of relaxation on the echo. Throughout both τ delays, spin–spin relaxation attenuates the phase-coherence created by the 90° pulse, and causes the transverse magnetization to decay with the rate constant T_2^{-1}. This dephasing, which results from the fluctuating magnetic fields arising from random molecular motions, is *not* refocussed by the 180° pulse. The NMR intensity of each line in the spin-echo spectrum is therefore given by

$$S(2\tau) = S(0)\exp(-2\tau/T_2).\tag{6.7}$$

The whole experiment is repeated with different τ delays, and T_2 is obtained from a plot of $\ln[S(2\tau)]$ against τ.

It will be clear from the above that the spin echo pulse sequence refocuses not only the dephasing arising from field inhomogeneity but also the chemical shift.

The phase angle $\Omega\tau$ accumulated during the first delay is exactly undone by the second delay, irrespective of the offset frequency, Ω. This feature makes $\tau-180°-\tau$ an important element in a large number of pulse sequences (e.g. Section 6.5).

6.5 INEPT

We now turn to a more complex NMR pulse sequence comprising five radiofrequency pulses (Fig. 6.14). The experiment is known by the acronym INEPT: Insensitive Nuclei Enhanced by Polarization Transfer. Its purpose is to enhance the weak NMR signals of a low-γ nucleus (e.g. ^{13}C or ^{15}N) by transferring polarization from a J-coupled high-γ nucleus (e.g. ^{1}H). It can be used on its own, as described below, or as part of a more complex pulse sequence (Section 6.6).

Consider a molecule containing two J-coupled spins, a proton and a carbon. The conventional ^{1}H and ^{13}C spectra are both doublets with lines at offset frequencies $\Omega_H\pm\pi J$ and $\Omega_C\pm\pi J$, respectively, where, as usual, the coupling constant J is in Hz. The two ^{1}H magnetization vectors are denoted $\mathbf{H}_\alpha$ and $\mathbf{H}_\beta$ according to the spin-state of the ^{13}C; $\mathbf{C}_\alpha$ and $\mathbf{C}_\beta$ are defined similarly in terms of the ^{1}H spin state.

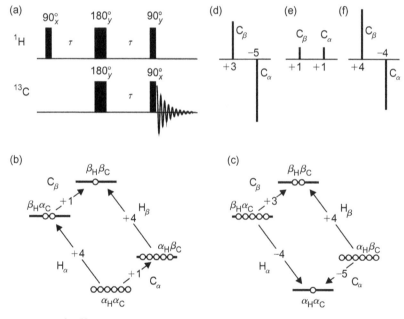

Fig. 6.14 The ^{1}H–^{13}C INEPT experiment. (a) Pulse sequence. (b–c) Energy levels and polarizations: at thermal equilibrium (b) and immediately after the $90°_y$(H) pulse but before the $90°_x$(C) pulse (c). The numbers on the arrows and the circles on the energy levels indicate the relative population differences of the four transitions. (c) is obtained from (b) by inverting the population differences as follows: 1. $\alpha_H\alpha_C \leftrightarrow \alpha_H\beta_C$, 2. $\beta_H\alpha_C \leftrightarrow \beta_H\beta_C$, 3. $\alpha_H\alpha_C \leftrightarrow \beta_H\alpha_C$. (d–f) ^{13}C spectra: resulting from (d) the INEPT pulse sequence, (e) following a single 90° pulse, and (f) the sum of the spectra in (d) and (e).

The pulse sequence comprises three pulses on the protons and two on the carbons (Fig. 6.14(a)). The delay τ is set equal to $(4J)^{-1}$: for a typical one-bond C–H coupling of ~150 Hz, $\tau = 1.67$ ms. Up to the end of the second τ delay, the ^{1}H pulse sequence $(90^{\circ}_x - \tau - 180^{\circ}_y - \tau -)$ is nothing more than a spin echo experiment (Fig. 6.13). Each of the ^{13}C pulses can be applied immediately before, at the same time as, or immediately after the corresponding ^{1}H pulse. In the following, we treat them as coming immediately after. At the end of the sequence, the ^{13}C free induction decay is recorded and Fourier transformed to give the ^{13}C spectrum.

Figure 6.15(a–g) shows how the magnetization vectors $\boldsymbol{H}_\alpha$ and $\boldsymbol{H}_\beta$ for the two ^{1}H lines behave during the pulse sequence. The first pulse, 90°_x(H), rotates both vectors from the $+z$-axis (a) to the $-y$-axis (b). They then start to precess in the xy-plane at their respective offset frequencies, $\Omega_H \pm \pi J$. Since the difference in these frequencies is $2\pi J$, $\boldsymbol{H}_\alpha$ and $\boldsymbol{H}_\beta$ accumulate a phase difference of $2\pi J \tau = 2\pi J/(4J) = \frac{1}{2}\pi \equiv 90°$ by the end of the first τ-delay (c). Next, the 180°_y(H) pulse flips $\boldsymbol{H}_\alpha$ and $\boldsymbol{H}_\beta$ around the y-axis (d). Then the 180°_y(C) pulse interchanges the α and β ^{13}C populations by inverting both lines in the ^{13}C spectrum. This has the effect of exchanging the $\boldsymbol{H}_\alpha$ and $\boldsymbol{H}_\beta$ labels (e). During the second τ-period, $\boldsymbol{H}_\alpha$ and $\boldsymbol{H}_\beta$ resume their precession in the xy-plane with the same frequencies as before so that by the end of the delay they are aligned along the $\pm x$ axes with a phase difference of 180° (f). This happens irrespective of the ^{1}H offset frequency (Ω_H) because the chemical shifts are refocussed by the spin-echo part of the pulse sequence. Next, the 180°_y(H) pulse rotates $\boldsymbol{H}_\alpha$ and $\boldsymbol{H}_\beta$ to the $\mp z$-axes,

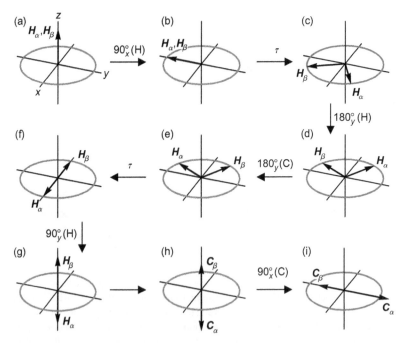

Fig. 6.15 The ^{1}H–^{13}C INEPT experiment. (a–g) ^{1}H magnetization vectors and (h–i) ^{13}C magnetization vectors at different stages during the pulse sequence.

respectively (g). The net effect of the pulse sequence up to this point is that $\boldsymbol{H}_\alpha$ has been rotated to the $-z$-axis while $\boldsymbol{H}_\beta$ has been returned to its starting position on the $+z$-axis. That is, the population difference for one of the two ^{1}H transitions has been inverted. In addition, the $180^\circ_y(C)$ pulse has inverted the population differences of both ^{13}C transitions.

Before considering the final $90^\circ_x(C)$ pulse, we need to look at the changes to the populations of the four states (Fig. 6.14(b,c)). At equilibrium, before the pulse sequence starts, the ^{1}H and ^{13}C polarizations are in the ratio of the magnetogyric ratios, i.e. $\gamma_H/\gamma_C \approx 4$. The intensities of both ^{13}C lines are equal to $+1$ in arbitrary units; in the same units the ^{1}H lines have intensity $+4$ (Fig. 6.14(b)). As described above, the first four pulses of the INEPT experiment have the net effect of inverting first the $\boldsymbol{C}_\alpha$ and $\boldsymbol{C}_\beta$ lines and then the $\boldsymbol{H}_\alpha$ line (Fig. 6.14(c)). The result is that the ^{13}C polarizations, which were both $+1$ at the beginning, are now $+3$ for the $\boldsymbol{C}_\beta$ line and -5 for the $\boldsymbol{C}_\alpha$ line.

We can now go back to the vector model (Fig. 6.15(h,i)). Immediately before the $90^\circ_x(C)$ pulse, as we have just seen, the ^{13}C magnetization vectors are arranged thus: $\boldsymbol{C}_\alpha$ along the $-z$-axis with length 5, and $\boldsymbol{C}_\beta$ along the $+z$-axis with length 3 (h). The final pulse rotates them onto the $\pm y$-axes (i). The ^{13}C spectrum, obtained by Fourier transforming the ensuing free induction decay, therefore contains an antiphase doublet in which the lines are $+3$ and -5 times the intensity of a conventional single-pulse spectrum (Fig. 6.14(d)).

The asymmetry of the ^{13}C doublet in the INEPT spectrum can be removed by adding a single-pulse ^{13}C spectrum (Fig. 6.14(e)) to give line intensities of ± 4 (Fig. 6.14(f)). Alternatively, the pulse sequence can be repeated with the phase of the $90^\circ_y(H)$ pulse inverted. This gives a spectrum with intensities -5 and $+3$ instead of $+3$ and -5. Subtracting the two spectra and dividing by two again gives an undistorted antiphase doublet with intensities $+4$ and -4 (Fig. 6.14(f)). Thus, polarization transfer from ^{1}H to ^{13}C enhances the lines in the ^{13}C spectrum by a factor $\gamma_H/\gamma_C \approx 4$. Nuclides with lower γ than ^{13}C have correspondingly larger INEPT enhancements.

At this point, we abandon the vector model. As our presentation of INEPT perhaps illustrates, it does not always give entirely clear insight into the operation of certain pulse sequence elements (e.g. the polarization transfer effected by the last two pulses) and there are plenty of NMR experiments where it is not applicable at all. Although beyond the scope of this book, there is a relatively simple and much more powerful way of understanding even quite complex experiments—the *product operator formalism* (Hore, Jones, and Wimperis (2015)).

6.6 Two-dimensional NMR

Principles

All the NMR spectra discussed so far have been *one-dimensional* in the sense that the NMR signal—the free induction decay, $s(t)$—was recorded as a function of a single time variable, t, and then Fourier transformed to give the spectrum

$S(\Omega)$ (eqn 6.4 and Fig. 6.9). However, a great many modern NMR experiments are *multi-dimensional*. Signals are recorded as a function of two or more time variables ($s(t_1, t_2, \cdots)$) so that, after multi-dimensional Fourier transformation, the spectra are functions of several frequency variables, $S(\Omega_1, \Omega_2, \cdots)$. In this section we outline how *two-dimensional* NMR experiments work and indicate the kinds of information they can provide.

We start with a molecule containing two weakly *J*-coupled protons, A and X. The simplest one-dimensional experiment (a 90° pulse followed by acquisition of the free induction decay, Fig. 6.1) would give the spectrum shown schematically in Fig. 6.16(a) with lines at offset frequencies $\Omega = \Omega_A \pm \pi J$ and $\Omega = \Omega_X \pm \pi J$. A two-dimensional spectrum of the same molecule might look something like Fig. 6.16(b). Instead of a single frequency variable, Ω, there are now two: Ω_1 and Ω_2. In this schematic spectrum all four frequencies are present in both dimensions, giving a total of 16 two-dimensional NMR lines arranged in groups of four. Two of these groups are centred on the line $\Omega_1 = \Omega_2$ and are called 'diagonal peaks'; they contain little information that is not present in a conventional one-dimensional spectrum (Fig. 6.16(a)). The other two groups are centred at different frequencies in the two dimensions and are known as 'cross peaks'. These are more interesting because their presence shows that the two spins from which they arise must be correlated in some way.

The information content of a two-dimensional spectrum is determined by the pulse sequence, a general form of which is shown in Fig. 6.16(c). The preparation

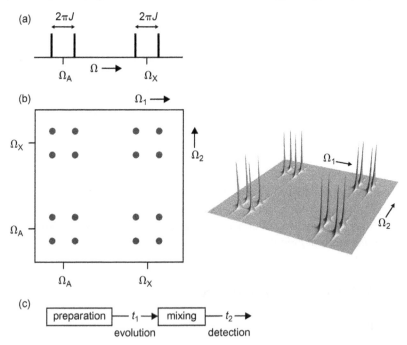

Fig. 6.16 Two-dimensional NMR of weakly coupled spin-$\frac{1}{2}$ nuclei, A and X. (a) One-dimensional spectrum. (b) Two representations of a schematic two-dimensional spectrum. (c) A schematic two-dimensional NMR pulse sequence.

and mixing periods each consist of one or more radiofrequency pulses. They are separated by a period of free precession, t_1 (the evolution period), and followed by detection of a free induction decay as a function of t_2. The sequence is repeated for different values of t_1 to build up a two-dimensional data set, $s(t_1, t_2)$. The general idea is that transverse magnetization of spin A, created by the preparation element, precesses at frequencies $\Omega_A \pm \pi J$ during t_1, is transferred to spin X at the mixing stage, and then precesses at frequencies $\Omega_X \pm \pi J$ during t_2. The result is a cross peak in the spectrum centred at $(\Omega_1, \Omega_2) = (\Omega_A, \Omega_X)$. In many cases there is also a cross peak centred at $(\Omega_1, \Omega_2) = (\Omega_X, \Omega_A)$ arising from magnetization transferred from X to A. In general, there are also diagonal peaks at $(\Omega_1, \Omega_2) = (\Omega_A, \Omega_A)$ and (Ω_X, Ω_X) arising from the magnetization that was not transferred between spins during the mixing period. See Hore, Jones, and Wimperis (2015) for more on the basics of two-dimensional NMR.

Fig. 6.17 Two-dimensional NMR pulse sequences: COSY and NOESY.

COSY and NOESY

NMR spectroscopists like acronyms.

Figure 6.17 shows the pulse sequences of two simple homonuclear two-dimensional NMR experiments: COSY (COrrelated SpectroscopY, pronounced 'cosy') and NOESY (Nuclear Overhauser Effect SpectroscopY, pronounced 'nosey'). Cross peaks in COSY spectra arise from spins with J-couplings, i.e. pairs of nuclei separated by up to three or four chemical bonds. NOESY cross peaks identify pairs of spins with significant dipolar coupling, i.e. nuclei less than ~5 Å apart in a molecule. In both cases, the first 90° pulse prepares the spins by rotating their magnetization vectors into the xy-plane where they evolve at their offset frequencies. In COSY, the second 90° (mixing) pulse transfers magnetization between J-coupled spins in much the same way as INEPT (Section 6.5). In NOESY, the $90°- \tau_m -90°$ section performs the mixing. During the τ_m delay (the 'mixing' time) z-magnetization of one of the spins (A) is transferred to z-magnetization of the other spin (X) via the NOE. The 90° pulses before and after τ_m are responsible for interconverting transverse and longitudinal magnetization, $\boldsymbol{M}_{x,y}(A) \to \boldsymbol{M}_z(A)$ and $\boldsymbol{M}_z(X) \to \boldsymbol{M}_{x,y}(X)$, respectively.

The object of a COSY experiment is to discover the network of J-couplings in a molecule, and so to arrive at spectral assignments—i.e. to determine which resonance corresponds to which nucleus. This process is illustrated in Figs 6.18 and 6.19.

Figure 6.18 shows the ^{11}B COSY spectrum of the borane $B_{10}H_{14}$. Cross peaks are only observed between directly bonded borons ($4 \leftrightarrow 1,2,3$ and $1 \leftrightarrow 3$) giving immediately an unambiguous assignment of peaks 2 and 4. Resonances 1 and 3, which arise, respectively, from two and four boron atoms, are easily distinguished by their 1:2 intensity ratio in a one-dimensional spectrum.

A slightly more complicated COSY spectrum is shown in Fig. 6.19. The nine CH protons in this sugar (a–i) give rise to ten cross peaks; all two-bond and three-bond spin–spin couplings are observed. Once again the pattern of scalar couplings is trivially deduced. The ambiguity in the assignment, arising from the two-fold symmetry of the connectivity pattern, can be resolved by chemical shift arguments or, more satisfactorily, by the observation (not shown) of NOEs from peak h to peaks d and f, and the absence of NOEs from b to g and e.

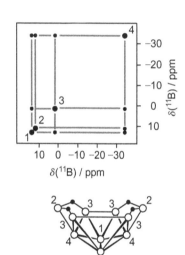

Fig. 6.18 Schematic ^{11}B COSY spectrum of $B_{10}H_{14}$. In the structure, the small solid circles represent bridging hydrogens; the ten terminal hydrogens (one per boron) are not shown. The four types of boron atom give rise to four cross peaks and hence four 'connectivities', shown below the structure. (After T. L. Venable, W. C. Hutton, and R. N. Grimes, *J. Am. Chem. Soc.*, **106** (1984) 29.)

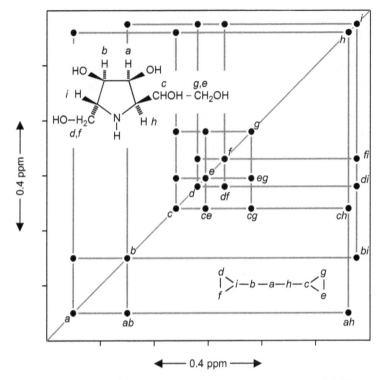

Fig. 6.19 Schematic 400 MHz ^{1}H COSY spectrum of the sugar shown at upper left (dissolved in D$_2$O). The CH protons are labelled *a–i*; the labile OH and NH protons are replaced by deuterons in D$_2$O and do not appear in the spectrum. The multiplet structure of the cross and diagonal peaks arising from *J*-coupling is not shown. The inset at lower right shows the network of *J*-couplings revealed by the pattern of cross peaks. Based on a spectrum kindly provided by M. R. Wormald and G. W. J. Fleet.

Figure 6.20 shows a NOESY spectrum of the protein lysozyme whose one-dimensional spectrum appeared at the beginning of Chapter 2. There are several hundred NOE cross peaks, each of which corresponds to a pair of protons separated by less than about 6 Å. As outlined in Section 5.8, the huge amount of information provided by such spectra can be used to deduce three-dimensional structures of molecules.

The NOESY technique can also be used to investigate *chemical exchange*, which, like the NOE, is capable of transferring magnetization from one site to another during the delay τ_m. As an example, Fig. 6.21 shows the spectrum of the heptamethylbenzene cation: the pattern of cross peaks (1–2, 2–3, 3–4) clearly reveals the existence of an intramolecular rearrangement in which a methyl group undergoes a series of 1,2 shifts.

The advantages of two-dimensional NMR may be seen by comparing NOESY with its one-dimensional equivalent—a series of one-dimensional ^{1}H{^{1}H} NOE experiments in which each resonance is saturated in turn: (a) crowded spectra are considerably simplified by spreading resonances into a second dimension; (b) there is no need for selective excitation of individual resonances, which is difficult or impossible when lines overlap; (c) measuring all NOEs simultaneously

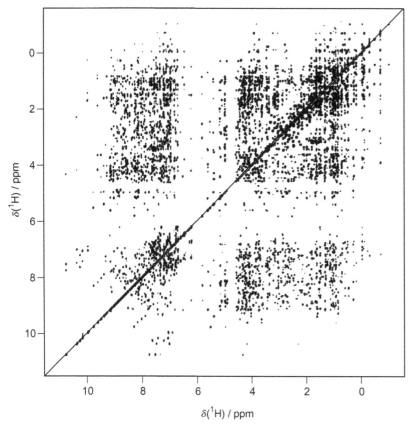

Fig. 6.20 950 MHz ^{1}H NOESY spectrum of the protein hen egg-white lysozyme in H_2O. Each of the off-diagonal 'spots' in this contour plot is an NOE cross peak. Artefacts arising from the solvent resonance at around 4.7 ppm have been removed. This spectrum was kindly provided by C. Redfield.

is more efficient than a series of one-dimensional experiments, in just the same way that FT NMR is superior to continuous wave NMR.

HSQC

COSY and NOESY are but two simple examples of the enormous range of powerful two-dimensional NMR experiments available to the chemist. Here, we outline one further experiment to give a flavour of what is possible. Much more on two-dimensional NMR can be found in, e.g., Freeman (1997a, 1997b, 2003), Levitt (2008), and Keeler (2010).

The HSQC (Heteronuclear Single Quantum Correlation) experiment (Fig. 6.22(a)) has the same function as COSY but correlates heteronuclear instead of homonuclear pairs of J-coupled spins, e.g. ^{1}H and ^{15}N. The pulse sequence comprises ten pulses and five delays, but its operation is easily understood. The first five pulses are simply an INEPT sequence whose purpose is to transfer ^{1}H polarization to ^{15}N transverse magnetization. As in Section 6.5, $\tau = (4J)^{-1}$. This is followed by the evolution period, t_1, in the middle of which is a 180° ^{1}H pulse

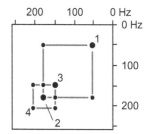

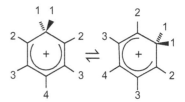

Fig. 6.21 A schematic ^{1}H two-dimensional spectrum of the four methyl resonances in the heptamethylbenzene cation obtained using the NOESY pulse sequence. The cross peaks indicate pairs of protons undergoing chemical exchange, in this case a 1,2 methyl shift. A different mechanism (e.g. a 1,3 shift) would lead to a different pattern of cross peaks. (After B. H. Meier and R. R. Ernst, *J. Am. Chem. Soc.*, **101** (1979) 6441.)

whose function is to refocus the ^{1}H chemical shifts and in this case the ^{1}H–^{15}N J-coupling too, in order that the ^{15}N spins precess at their chemical shift offset frequency, Ω_N, during t_1. The final four pulses are the reverse of an INEPT sequence (minus the first pulse); their purpose is to transfer the ^{15}N magnetization back to the protons so that the ^{1}H free induction decay can be recorded. If the ^{15}N spins are decoupled during t_2, the result is a single cross peak at $(\Omega_1, \Omega_2) = (\Omega_N, \Omega_H)$. The sensitivity of this experiment is high because the starting point is the large (relative to ^{15}N) ^{1}H thermal equilibrium polarization and the end point is the ^{1}H free induction decay which, being at a higher frequency, can be more sensitively detected than the ^{15}N signal.

Figure 6.22(b) shows the ^{1}H–^{15}N HSQC spectrum of ^{15}N-enriched lysozyme in H$_2$O. It contains a single peak from the backbone amide NH of every amino acid

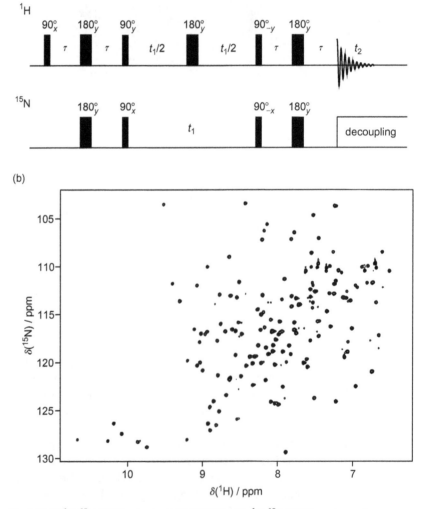

Fig. 6.22 (a) ^{1}H–^{15}N HSQC pulse sequence. (b) 950 MHz ^{1}H–^{15}N HSQC spectrum of ^{15}N-enriched hen egg-white lysozyme in H$_2$O. This spectrum was kindly provided by C. Redfield.

residue, of which there are 129 in this protein, except the prolines which do not have backbone NH groups. In addition, there are two peaks from the amide NH_2 of each of the asparagine and glutamine residues and one from the indole NH of each of the six tryptophan residues (in the lower left corner).

6.7 Three-dimensional NMR

Although two-dimensional NMR techniques have enormous advantages over one-dimensional methods, they run into problems with large molecules such as proteins. As the molecular mass increases, rotational tumbling becomes slower and spin–spin relaxation times shorter. The NMR lines consequently become broader and overlap more strongly with one another even when spread out into two dimensions. A solution is to add a third dimension, i.e. to record data as a function of three time-variables using the pulse sequence structure shown in Fig. 6.23. Compared to Fig. 6.16(c), there are now two evolution periods, t_1 and t_2, each followed by a mixing period, and finally a detection period, t_3. This can be regarded as a combination of two two-dimensional pulse sequences with the detection period of the first removed and the preparation period of the second replaced by the mixing period of the first (Fig. 6.23).

Consider, as an example, the combination of a ^{1}H–^{1}H NOESY experiment followed by a ^{1}H–^{15}N HSQC experiment (known, not surprisingly, as NOESY-HSQC) applied to a molecule containing two protons H_A and H_X which are less than ~5 Å apart, one of which (H_X) is directly bonded, and therefore J-coupled, to a nitrogen, N_X. In the preparation period, the equilibrium polarization of H_A is rotated into the xy-plane where it precesses at offset frequency Ω_H^A during t_1. The first mixing period transfers magnetization from H_A to H_X via the NOE and thence to N_X by an INEPT sequence. The N_X magnetization precesses during t_2 at frequency Ω_N^X and is then transferred back to H_X by a reverse INEPT. Finally, H_X magnetization, at frequency Ω_H^X, is detected during the t_3 period. The result, after the data have been Fourier transformed in all three dimensions is a three-dimensional spectrum containing a peak with coordinates $(\Omega_1, \Omega_2, \Omega_3) = \left(\Omega_H^A, \Omega_N^X, \Omega_H^X\right)$ demonstrating that H_A and H_X are close in space and that H_X is bonded to N_X. For a molecule with many protons and nitrogens (e.g. a protein), a two-dimensional

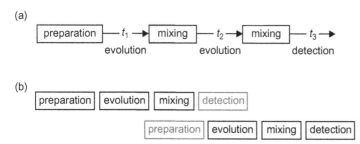

Fig. 6.23 (a) A schematic three-dimensional NMR pulse sequence. (b) Its construction from two two-dimensional NMR experiments.

slice through such a spectrum at the Ω_2 frequency of one of the nitrogens contains all the ^{1}H–^{1}H NOE cross peaks of the proton bound to that nitrogen. For more on three-dimensional NMR see e.g. Cavanagh et al. (2007).

6.8 Summary

- The vector model provides a simple picture of some elementary NMR experiments.
- It can be used to describe the effects of radiofrequency pulses, periods of free precession, and spin relaxation.
- The rotating frame is a crucial simplifying feature of the vector model.
- NMR spectra are obtained from free induction decays by Fourier transformation.
- Spin–spin relaxation is the loss of phase-coherence amongst the spins.
- T_1 and T_2 can be measured using inversion recovery and spin-echo pulse sequences, respectively.
- Polarization transfer (INEPT) from high-γ spins can be used to enhance the sensitivity of low-γ spins.
- Two-dimensional NMR spectra are obtained by measuring signals as a function of two time-variables.
- The two-dimensional experiments COSY and NOESY expose correlations between homonuclear pairs of spins.
- HSQC reveals correlations between ^{13}C or ^{15}N nuclei and directly bonded ^{1}H nuclei.
- Two-dimensional NMR experiments can be combined to produce three-dimensional NMR experiments.

6.9 Exercises

Answers to the exercises are provided at the back of the book. Full worked solutions are available on the Online Resource Centre at www.oxfordtextbooks.co.uk/orc/hore_nmr2e/

1. What is the duration of a ^{1}H 90° pulse when $B_1 = 1.00$ mT?

2. How long does it take for the free induction decay of an NMR line with $T_2 = 0.5$ s to decay to 1% of its initial amplitude?

3. The populations of the energy levels of a collection of spin-$\frac{1}{2}$ nuclei at thermal equilibrium are $n_\alpha = n_\alpha^{eq}$ and $n_\beta = n_\beta^{eq}$. In terms of n_α^{eq} and n_β^{eq}, what are the two populations after (a) a 90° pulse and (b) a 180° pulse?

4. When $B_1 \gg \Delta B$, the effective field during a radiofrequency pulse, $\boldsymbol{B}_{eff}$, lies in the xy-plane of the rotating frame. (a) On a 400 MHz spectrometer with a 10 μs ^{1}H 90° pulse, over what range of offset frequencies Ω is the angle between $\boldsymbol{B}_{eff}$ and the xy-plane less than 10°? (b) Taking $\delta = 5$ ppm at $\Omega = 0$, what is the corresponding range of chemical shifts?

5. The NMR spectrum obtained by Fourier transforming a free induction decay of the form $A\cos(\Omega_0 t)\exp(-t/T_2)$ comprises a single line at offset frequency $\Omega = \Omega_0$ with amplitude A. Describe the appearance of the spectrum that would be obtained from the free induction decay $A\cos(\Omega_0 t)\cos(\pi J t)\exp(-t/T_2)$.

6. How many free induction decays would it be necessary to add together to improve the signal-to-noise ratio by a factor of 50?

7. The following data were measured in an inversion recovery experiment. Determine the value of T_1.

τ/s	2.0	5.0	10.0	20.0	100.0
$S(\tau)$	−0.702	−0.336	+0.107	+0.601	+1.000

8. The spin echo pulse sequence in Fig. 6.13 refocusses magnetization along the $-y$-axis. What would happen if the phase of the 180° pulse were changed from y to x?

9. Predict which cross peaks would have been seen in the two-dimensional spectrum of the heptamethylbenzene cation (Fig. 6.21) if the rearrangement of the methyl group had been (a) 1,3 shifts, (b) 1,4 shifts, (c) random shifts.

10. The ^1H COSY spectrum of a di-substituted benzene, C_6H_4XY, contains three cross peaks on each side of the diagonal. Considering only three-bond J-couplings determine whether X and Y, which do not contain magnetic nuclei, are *ortho*, *meta*, or *para*.

Appendix A. Magnetic dipoles

Magnetic dipoles and the magnetic fields they generate crop up in different contexts in different chapters of this book. To avoid excessive repetition, some basic properties are summarized here.

Consider a classical microscopic magnetic moment μ (a 'point dipole'), placed at the origin of an (x,y,z) coordinate system, with magnitude μ and pointing along the positive z-axis (Fig. A.1). It produces a magnetic field $\boldsymbol{B}_{\text{dip}}$ that has cylindrical symmetry around the z-axis. At a point in the xz-plane specified by the polar coordinates r and θ (i.e. $x = r\sin\theta$, $y = 0$, $z = r\cos\theta$) the x, y, and z components of $\boldsymbol{B}_{\text{dip}}$ are given by

$$B_{\text{dip},x} = \left(\frac{\mu_0}{4\pi}\right)\left(\frac{\mu}{r^3}\right)(3\sin\theta\cos\theta)$$

$$B_{\text{dip},y} = 0 \qquad\qquad (A.1)$$

$$B_{\text{dip},z} = \left(\frac{\mu_0}{4\pi}\right)\left(\frac{\mu}{r^3}\right)(3\cos^2\theta - 1)$$

where $\mu_0 = 4\pi\times10^{-7}$ H m^{-1} is the vacuum permeability. The magnitude of the field falls off as the *inverse cube* (r^{-3}) of the distance from μ. $\boldsymbol{B}_{\text{dip}}$ has the same form as the magnetic field produced by a microscopic bar magnet or loop of wire carrying a steady current.

There are various ways of depicting the vector field $\boldsymbol{B}_{\text{dip}}$. The most common is a plot of the magnetic field lines (Fig. A.2(a)) obtained by joining end-to-end infinitesimal vectors representing $\boldsymbol{B}_{\text{dip}}$ (so that $\boldsymbol{B}_{\text{dip}}$ is everywhere tangent to a field line). Although this diagram gives little idea of the magnitude of the dipolar field, it does show its *direction* quite nicely. On the z-axis ($\theta = 0$ or $180°$), $\boldsymbol{B}_{\text{dip}}$ is parallel to μ; on the x-axis (or indeed anywhere in the $z = 0$ plane) $\boldsymbol{B}_{\text{dip}}$ is antiparallel to μ, i.e. it points parallel to the *negative z-axis*. At the 'magic angle' positions where $3\cos^2\theta = 1$ (i.e. $\theta \approx \pm54.7°$ and $\theta \approx 180 \pm 54.7°$) the z-component of the field vanishes, and $\boldsymbol{B}_{\text{dip}}$ is perpendicular to μ. These properties can of course also be seen directly from eqn A.1. Thus the direction of $\boldsymbol{B}_{\text{dip}}$ depends on θ, while its magnitude is determined mainly by r, through the r^{-3} distance-dependence. A different way of visualizing the z-component of the dipolar field is by means of a density plot (Fig. A.2(b)), which shows the variation of $B_{\text{dip},z}$ with (r,θ). Note that if the magnetic moment μ is rotated, its magnetic field rotates with it. For example, $\boldsymbol{B}_{\text{dip}}$ is inverted if μ points along the negative z-axis.

Now consider the interaction between two classical magnetic dipoles, μ_A and μ_X, both pointing parallel to the positive z-axis, with the first at the origin and the

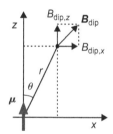

Fig. A.1 Parameters associated with the magnetic field generated by a magnetic dipole μ. The components of $\boldsymbol{B}_{\text{dip}}$ at a position in the xz-plane with polar coordinates (r,θ) are given in eqn A.1.

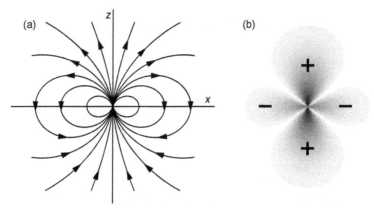

Fig. A.2 Representations of the dipolar magnetic field $\boldsymbol{B}_{dip}$ generated by a magnetic dipole $\boldsymbol{\mu}$ pointing along the positive z-axis: (a) magnetic field lines; (b) the z-component of $\boldsymbol{B}_{dip}$. Regions in which $\boldsymbol{B}_{dip,z}$ is positive and negative are labelled + and −, respectively. The angular part of $\boldsymbol{B}_{dip,z}$ (i.e. $3\cos^2\theta - 1$) is identical to that of the d_{z^2} orbital in atomic hydrogen.

second at (r,θ). The dipolar field produced by $\boldsymbol{\mu}_A$ at the position of $\boldsymbol{\mu}_X$ is given by eqn A.1 with $\boldsymbol{\mu}$ replaced by $\boldsymbol{\mu}_A$. The interaction energy is

$$E = -\boldsymbol{\mu}_X \cdot \boldsymbol{B}_{dip} = -\mu_X B_{dip,z} = -\left(\frac{\mu_0}{4\pi}\right)\left(\frac{\mu_A \mu_X}{r^3}\right)\left(3\cos^2\theta - 1\right). \tag{A.2}$$

The interaction depends on the two magnetic moments, their separation, and the angle, θ, between the z-axis and the internuclear vector. E is negative in the regions of Fig. A.2(b) labelled +, positive in the regions labelled −, and zero at the magic angles where $3\cos^2\theta = 1$.

Dipolar interactions are discussed in the context of chemical shifts (Section 2.4), spin–spin coupling (Sections 3.6 and 3.8), and spin relaxation (Chapter 5).

Glossary

Chemical exchange. Exchange of nuclei between sites with different NMR frequencies as a result of a dynamic chemical or conformational equilibrium.

Chemical shift. Dependence of the NMR frequency of a nucleus in a molecule on the local electronic structure.

COSY. Homonuclear two-dimensional NMR experiment that correlates spins via their mutual J-couplings (COrrelated SpectroscopY).

Cross relaxation. Concerted spin relaxation of pairs of dipolar-coupled nuclei; responsible for the nuclear Overhauser effect.

Delay. Period of free precession in an NMR pulse sequence.

Dipolar coupling. Through-space magnetic dipole–dipole interaction of one nuclear spin with another.

Electric quadrupole moment. Quantity that determines the strength of the interaction of a nuclear spin with an electric field gradient.

Equivalent spins. Nuclei with identical chemical shifts, usually as a consequence of molecular symmetry.

Fast exchange. Chemical exchange regime in which the relevant rate constants are large compared to the difference in NMR frequencies of the exchanging spins.

Fermi contact interaction. Isotropic magnetic interaction of an electron and a nucleus that gives rise to J-coupling.

Fourier transform. Mathematical operation used to convert a (time-domain) free induction decay into a (frequency-domain) NMR spectrum.

Free induction decay. Time-domain NMR signal produced by a pulse sequence.

Free precession. Motion of the magnetization vector of an NMR sample in the absence of radiofrequency fields.

Heteronuclear spins. Nuclei of different nuclide types.

Homonuclear spins. Nuclei of the same nuclide type.

INEPT. Heteronuclear pulse sequence used to enhance the NMR signal of a low-γ nucleus (Insensitive Nuclei Enhanced by Polarization Transfer).

J-coupling. Through-bond magnetic interaction of one nuclear spin with another.

Larmor frequency. NMR precession frequency of a nuclear spin in a magnetic field.

Magnetic quantum number. Describes the component of spin angular momentum along the magnetic field axis.

Magnetization. Net magnetic moment of an NMR sample.

Magnetogyric ratio (γ). Quantity that relates the magnetic moment of a nucleus to its spin angular momentum.

Multiplets. Spectral features (doublets, triplets, quartets, ...) arising from J-coupling.

NMR timescale. Reciprocal of the difference in NMR frequencies of spins undergoing chemical exchange.

NOESY. Two-dimensional NMR experiment that correlates spins via their mutual dipolar couplings (Nuclear Overhauser Effect Spectros-copY).

Nuclear magnetic moment. Quantity that determines the strength of the interaction of a nuclear spin with a magnetic field.

Nuclear Overhauser effect (NOE). Transfer of polarization from one nuclear spin to another via dipolar cross relaxation.

Polarization. Alignment of nuclear spins.

Pulse sequence. Collection of radiofrequency pulses and delays designed to elicit a specific response from an NMR sample.

Quadrupolar relaxation. Spin relaxation arising from the interaction of the quadrupole moment of a nucleus with an electric field gradient in a molecule.

Radiofrequency pulse. Short burst of monochromatic radiofrequency radiation used to excite spins in an NMR experiment.

Rotating frame. Coordinate system in which the magnetic field of a radiofrequency pulse appears stationary.

Rotational correlation time. Average time a molecule in a liquid takes to rotate through approximately one radian.

Shielding. Change in the net magnetic field experienced by a nucleus as a result of local magnetic fields produced by induced electronic currents in molecules; origin of the chemical shift.

Slow exchange. Chemical exchange regime in which the relevant rate constants are small compared to the difference in NMR frequencies of the exchanging spins.

Spin. Spin angular momentum. Also a synonym for 'magnetic nucleus'.

Spin-I. Nucleus with spin quantum number I.

Spin angular momentum. Fundamental property of particles such as protons and neutrons, and therefore nuclei, that gives rise to nuclear magnetism.

Spin echo. NMR signal arising from the refocussing effect of a 180° pulse.

Spin–lattice relaxation time, T_1. Characteristic time for the relaxation of longitudinal (z) magnetization.

Spin quantum number. Quantifies the magnitude of the spin angular momentum of a nuclide.

Spin relaxation. Process by which nuclear spins come to thermal equilibrium.

Spin–spin coupling. Magnetic interaction of one nuclear spin with another.

Spin–spin relaxation time, T_2. Characteristic time for the relaxation of transverse (xy) magnetization.

Strongly coupled spins. Nuclei whose mutual J-coupling is comparable to the difference in their chemical shift frequencies.

Two-dimensional/three-dimensional NMR. Experiments in which NMR spectra are recorded as a function of two/three frequency variables.

Vector model. Simple pictorial way of understanding the behaviour of nuclear magnetization in an NMR experiment.

Weakly coupled spins. Nuclei whose mutual J-coupling is much smaller than the difference in their chemical shift frequencies.

Bibliography

Apperley, D., Harris, R. K., and Hodgkinson, P. (2012). *Solid state NMR: basic principles and practice*. Momentum Press, New York.

Atkins, P. and Friedman, R. (2011). *Molecular quantum mechanics* (5th edn). Oxford University Press, Oxford.

Berger, S. and Braun, S. (2004). *200 and more NMR experiments* (2nd edn). Wiley-VCH, Weinheim.

Bonhomme, C., Gervais, C., Babonneau, F., Coelho, C., Pourpoint, F., Azaïs, T., Ashbrook, S. E., Griffin, J. M., Yates, J. R., Mauri, F., and Pickard, C. J. (2012). First-principles calculation of NMR parameters using the gauge including projector augmented wave method: a chemist's point of view, *Chem. Rev.* **112**, 5733–79.

Bovey, F. A. (1988). *Nuclear magnetic resonance spectroscopy* (2nd edn). Academic Press, San Diego.

Breitmaier, E. (1993). *Structure elucidation by NMR in organic chemistry*. Wiley, Chichester.

Campbell, I. D. (2012). *Biophysical techniques*. Oxford University Press, Oxford.

Carrington, A. and McLachlan, A. D. (1967). *Introduction to magnetic resonance*. Harper and Row, New York.

Cavanagh, J., Fairbrother, W. J., Palmer, A. G., Rance, M., and Skelton, N. J. (2007). *Protein NMR spectroscopy* (2nd edn). Elsevier Academic Press, Burlington.

Claridge, T. D. W. (2009). *High-resolution NMR techniques in organic chemistry*. Elsevier, Oxford.

Duer, M. J. (2004). *Introduction to solid-state NMR spectroscopy*. Blackwell, Oxford.

Ernst, R. R., Bodenhausen, G., and Wokaun, A. (1987). *Principles of nuclear magnetic resonance in one and two dimensions*. Clarendon Press, Oxford.

Freeman, R. (1997a). *A handbook of nuclear magnetic resonance* (2nd edn). Addison Wesley Longman, Harlow.

Freeman, R. (1997b). *Spin choreography*. Spektrum, Oxford.

Freeman, R. (2003). *Magnetic resonance in chemistry and medicine*. Oxford University Press, Oxford.

Friebolin, H. (2011). *Basic one-and two-dimensional NMR spectroscopy* (5th edn). Wiley-VCH, Weinheim.

Goldman, M. (1988). *Quantum description of high-resolution NMR in liquids*. Clarendon Press, Oxford.

Günther, H. (2013). *NMR spectroscopy* (3rd edn). Wiley-VCH, Weinheim.

Harris, R. K. (1983). *Nuclear magnetic resonance spectroscopy*. Pitman, London.

Hore, P. J., Jones, J. A., and Wimperis, S. (2015). *NMR: the toolkit. How pulse sequences work* (2nd edn), Oxford University Press, Oxford.

Keeler, J. (2010). *Understanding NMR spectroscopy* (2nd edn). Wiley, Chichester.

Kwan, A. H., Mobli, M., Gooley, P. R., King, G. F., and Mackay, J. P. (2011). Macromolecular NMR spectroscopy for the non-spectroscopist, *FEBS J.* **278**, 687–703.

Levitt, M. H. (2008). *Spin dynamics* (2nd edn). Wiley, Chichester.

Munowitz, M. (1988). *Coherence and NMR*. Wiley, New York.

Neuhaus, D. and Williamson, M. (1989). *The nuclear Overhauser effect in structural and conformational analysis*. VCH Publishers, New York.

Pregosin, P. S. (2002). *NMR in organometallic chemistry*. Wiley-VCH, Weinheim.

Sanders, J. K. M. and Hunter, B. K. (1993). *Modern NMR spectroscopy: a guide for chemists* (2nd edn). Oxford University Press, Oxford.

Slichter, C. P. (1990). *Principles of magnetic resonance* (3rd edn). Springer-Verlag, Berlin.

Wehrli, F. W., Marchand, A. P., and Wehrli, S. (1988). *Interpretation of carbon-13 NMR spectra* (2nd edn). Wiley, Chichester.

Williams, D. H. and Fleming, I. (2007). *Spectroscopic methods in organic chemistry* (6th edn). McGraw-Hill, London.

Wüthrich, K. (1986). *NMR of proteins and nucleic acids*. Wiley, New York.

Answers to exercises

Chapter 1

1. $I = 0$: $^{32}_{16}S$ and $^{40}_{20}Ca$. $I = \frac{1}{2}$: $^{57}_{26}Fe$ and $^{119}_{50}Sn$. $I = 1$: $^{6}_{3}Li$ and $^{14}_{7}N$.
2. $\frac{3}{2}$.
3. $2.884 \times 10^{-27}\,J\,T^{-1}$.
4. (a) 1000.0 MHz. (b) 251.5 MHz. (c) 101.4 MHz.
5. 2.129 kHz.
6. ^{15}N.
8. 5.992×10^{-5}.
9. (a) 125 GHz. (b) 0.96 K.
10. 1.22×10^4.

Chapter 2

1. 3.675 kHz.
2. 5.9 ppm.
3. ~34.3 kHz.
4. (a) 1,4-dichlorobenzene. (b) 2,2-dichloropropane.
   (c) 1,1,1,3,3,3-hexachloropropane or 1,1,2,2,3,3-hexachloropropane.
   (d) 1,1-dichlorocyclopropane. (e) p-benzoquinone.
5. (a) 3, 4, and 2. (b) 1 (1,1-dichlorocyclopropane), 2 (trans-
   1,2-dichlorocyclopropane), and 3 (cis-1,2-dichlorocyclopropane).
6. (a) Isomer 1: $CCl_3–CCl_2–CH_2Cl$. Isomer 2: $CCl_3–CHCl–CHCl_2$. (b) Both have
   one ^1H chemical shift and two ^{13}C chemical shifts.
7. sp^2 (alkenes).
8. $CHBr_3$ highest. CH_3Br lowest.
9. Meta highest. Ortho lowest.
10. 315 pm.

Chapter 3

1. (a) T-shaped. (b) Square pyramid. (c) 2-Chloropropane. (d) Chloroethylene
   oxide. (e) Square pyramid; $I = \frac{7}{2}$.
2. (a) 1. (b) 12. (c) 1. (d) 4. (e) 1.
3. (a) 1:2:3:2:1. (b) 7.
4. (a) 2-chlorobutane. (b) 1-chlorobutane.
5. Singlet, doublet, doublet, and triplet.
6. 2.8 ppm, 7.4 ppm, 6 Hz.
7. (a) Magnetically. (b) Chemically. (c) Magnetically. (d) Chemically.

8.   (a) 1.09. (b) 3.30.

9.   0.832 *gauche*, 0.168 *trans*.

10. (a) 152.6 pm. (b) 146.7 pm.

Chapter 4

1.   (a) 31.8 Hz. (b) 40.2 Hz.

2.   (a) $4.4 \, s^{-1}$; $18.9 \, s^{-1}$. (b) $88.7 \, kJ \, mol^{-1}$; $1.15 \times 10^{13} \, s^{-1}$.

3.   (a) two lines; (b) lines just merged into one; (c) one line.

4. $10.6 \, kJ \, mol^{-1}$.

5. $p_A = 0.995$, $p_B = 0.005$.

6.   100 Hz.

7. $1.0 \times 10^8 \, dm^3 \, mol^{-1} \, s^{-1}$.

8.   5.19 ppm.

9.   4 (fast rotation around N–benzene bond).

10. 7.21.

Chapter 5

1.   3.18 s.

2. $4.61 \, T_1$.

3.   (a) 318 ps. (b) 0.70 s.

4.   (a) 109 pm. (b) Longer.

5.   (a) H5. (b) H2.

6.   The protein.

7.   1.13 ns

8.   No change when $\omega_0 \tau_c \ll 1$. T_1 is increased by a factor of four when $\omega_0 \tau_c \gg 1$. The minimum T_1 is larger by a factor of two and occurs at half the value of τ_c.

10. 297 ps.

Chapter 6

1.   5.87 µs.

2.   2.30 s.

3.   (a) $n_\alpha = n_\beta = \frac{1}{2}\left(n_\alpha^{eq} + n_\beta^{eq}\right)$. (b) $n_\alpha = n_\beta^{eq}$, $n_\beta = n_\alpha^{eq}$.

4.   (a) $-27{,}700 < \Omega < +27{,}700 \, rad \, s^{-1}$. (b) $+16 > \delta > -6.0$ ppm.

5.   Two lines with frequencies $\Omega_0 \pm \pi J$ (i.e. a doublet), each with amplitude $\frac{1}{2} A$.

6.   2500.

7.   12.4 s.

8.   Refocussing along +y axis.

9.   (a) 1–3 and 2–4. (b) 1–4 and 2–3. (c) 1–2, 1–3, 1–4, 2–3, 2–4, and 3–4.

10. *ortho*.

Index